The Materiality of Interaction

The Materiality of Interaction

Notes on the Materials of Interaction Design

Mikael Wiberg

The MIT Press
Cambridge, Massachusetts
London, England

This book was set in ITC Stone Serif Std by Westchester Publishing Services. Printed and bound in the United States of America.

Library of Congress Cataloging-in-Publication Data
Names: Wiberg, Mikael, 1974- author.
Title: The materiality of interaction : notes on the materials of interaction
    design / Mikael Wiberg.
Description: Cambridge, MA : The MIT Press, [2018] | Includes bibliographical
    references and index.
Identifiers: LCCN 2017025621 | ISBN 9780262037518 (hardcover : alk. paper)
Subjects: LCSH: Human-computer interaction. | Materials--Computer simulation.
    | Human-machine systems.
Classification: LCC QA76.9.H85 W486 2018 | DDC 004.01/9--dc23 LC record available
    at https://lccn.loc.gov/2017025621

10   9   8   7   6   5   4   3   2   1

Contents

Acknowledgments

In this book I suggest that interaction design is ultimately a relational concern. Further on, I describe how the materials at play in such design processes stand in close relation to a core aspect of any interaction design project— that is, *interaction*. This concept also serves as a good description of the process behind the creation of this book. For sure, the writing of this book has been a relational process in that many of the ideas presented here were formed through a great number of discussions with a great number of great people in our field. For all of those discussions, I am so grateful! It has been a true pleasure to talk to each of you, and I cannot thank you all enough for the conversations we have had over the last few years.

To mention a few of the people to whom I am particularly thankful, I would first of all like to thank Doug Sery, Senior Acquisitions Editor at the MIT Press, for the discussions we have had about the scope and the focus of this book. I have really enjoyed working with you throughout the whole project! This project started a few years ago when we met up during the DRS'14 conference, and I am so thankful for the opportunity you gave me to start it. I am also thankful for the support from Professors Erik Stolterman, Ken Friedman, and Pelle Ehn, who early on showed their support and enthusiasm for this project. Your support has been very important, as it has inspired me to explore this topic from a design theory point of view, and I cannot thank you enough for your support and interest in this project! Thank you!

I would also like to express my gratitude to a number of people in our community. Without our close collaborations and in-depth discussions, this book would not exist; that I know for sure. First of all, I would like to thank Erica Robles-Anderson. I truly enjoyed working with you on the ideas that we developed over a number of coauthored papers; some of these ideas, as you can see, have also found their way into this manuscript. In addition, I would like to thank all of you who were part of the "Material Interactions" panel that we organized at the CHI conference in 2012, including Prof. Hiroshi Ishii, Prof. Paul Dourish, Petra Sundström, Daniela Rosner, Anna Vallgårda, Tobie Kerridge, and Mark Rolston. Great thanks to all of you! The discussions that we had before, during, and after our panel at CHI have motivated me to dig deeper into these questions concerning *the materiality of interaction*. You are all a great source of inspiration to me, and I am so grateful for the collaborations and the discussions we have had! I would also like to thank Jofish Kaye and Prof. Peter Thomas for our collaboration in organizing a special issue of *Personal and Ubiquitous Computing* on material interactions in 2013. That collaboration pushed forward my thinking on the materiality of interaction. I would also like to thank Heekyoung Jung for our collaboration on metaphors and materiality, which has indeed influenced my thinking on this topic. My deep thanks go as well to Prof. Elisa Giaccardi and Prof. Ron Wakkary for our truly inspiring discussions on interaction design that seeks to manifest the digital in material form. The work you do is a great source of inspiration for me! In addition, I would like to thank Prof. Jeffery Bardzell, Shaowen Bardzell, and Shad Gross for the discussions we have had on materiality in relation to interaction design in general and on the notion of non-skeuomorpic design in relation to materiality in particular. Our discussions motivated me to dig deeper into this issue, which became an important part of this book.

I would also like to thank a number of people in my own department, the Department of Informatics at Umeå University in Sweden, with whom I have collaborated over the years and who have been extremely valuable peers and sources of inspiration. Here I would in particular like to thank Andreas Lund, Fatemeh Moradi, Maliheh Ghajargar, Prof. Daniel Fällman, and Prof. Charlotte Wiberg. Great thanks to all of you!

My thanks go as well to the people who served as the anonymous reviewers for the MIT Press on earlier versions of this manuscript. Your feedback has been extremely valuable for me, and your input has indeed helped me to improve and further develop this manuscript. I am so grateful for your close readings and for all of the constructive comments on earlier versions of this book!

Finally, I would like to thank Johan Bodén for creating the graphical illustrations for this book. You have a great feeling for the details and I cannot thank you enough for bringing all the pictures and illustrations in this book together in a unifying graphical format. In closing, I would also like to thank Susan Buckley, Matthew Abbate, and the rest of the staff at the MIT Press for making this book project come true! Great thanks!

You are all amazing!

Introduction

Computing is increasingly intertwined with our physical world. From smart watches to connected cars, to the Internet of Things and 3D printing, the trend toward combining digital and analog materials in design is no longer an exception, but a hallmark for where interaction design is going in general. Computational processing increasingly involves physical materials, computing is increasingly manifested and expressed in physical form, and interaction with these new forms of computing is increasingly mediated via physical materials. Interaction design is therefore increasingly a material concern. Welcome to a book on the materiality of interaction, and material-centered interaction design!

In this book, I discuss the contemporary trend in interaction design toward material interactions. After describing the ways in which interaction design is increasingly about materials, I propose "material-centered interaction design" as a method for working with materials in interaction design projects.

Interaction design is the design of interaction through material configurations, and as such, it is increasingly about the ways we can configure materials to enable interaction with and through materials. Sometimes we refer to such material compositions as "interactive systems," while in other cases we talk about "smart objects," or even the "Internet of Things." Still, it is the integration, the configuration, and the composition of a wide range of materials that holds these interactive designs together. It is our ability to imagine such interactive compositions as well as our ability to reimagine what traditional materials can do from the viewpoint of interaction design that defines the design space for interaction design. A narrow focus on interaction design would be, for example, to think about what the computer can do and how we can design a user interface for that computational machine. Far beyond such a narrow focus, I suggest that since computing can take any form, and since any material can be reactivated and accordingly reimagined in a computational moment, then interaction design becomes the practice of imagining interaction design in new forms, and enabled through new material compositions. In short, and with this as a point of departure, interaction design becomes the practice of imagining and designing interaction through material manifestations; and in return, this practice is ultimately a concern for the materiality of interaction.

In an attempt to contribute to the current development, this book offers a *material-centered approach to interaction design* as a fundamental design method for working across digital, physical, and even "immaterial" materials in interaction design projects.

As this book will both demonstrate and theorize, the materiality of interaction design is not restricted to these digital and physical materials. Far beyond that, I propose that *interaction is the core materiality and the focus of interaction design*. In any interaction design project, we always need to envision a particular form of interaction, a particular turn-taking or other form of interplay between users and computers, between people interacting via computers, or even for machine-to-machine interaction. How we form that interaction, that is, how we manifest interaction in materials, is an eternal and expanding question for interaction design as a field. As new computational opportunities, new materials, and new interaction modalities are constantly being developed, we are constantly expanding the design space of interaction design even further.

As we move through different technological trends and different phases of developments, our palette of ways we can manifest ideas about interaction in computational materials is increasingly expanding. Not only do the development of new smart materials, augmented reality (AR), and the Internet of Things (IoT) bring new opportunities for interaction design across the physical and digital landscape, but all these opportunities also call for new compositional skills. Today's interaction designers need to know about existing frameworks, code libraries, materials, and components, and they need to have compositional skills to bring such elements together into computational wholes. This book is about that need, and it is about how the materiality of interaction is dependent on the particular ways such wholes are formed—whether the materiality is built up around metaphors or the materiality of interaction is more literally defined in physical form.

Although it might nowadays seem natural to think about computing in material terms (at least from the viewpoint of the growing interest in the Internet of Things, tangible computing, and embedded systems), it has in fact not always been the case. On the contrary, the area of interaction design has departed from a much more immaterial ground. Contrary to a "material ground," it is actually design with the "immaterial," and the design of "representations" and "symbols," not physical "materials," that has been our origin story. By making observations of the world (typically through user studies) and then creating representations of some aspects of that world, we have built a profession focused on how to get a computer to process these representations and how to configure computing machinery to display these representations to enable us to see and manipulate information—i.e., to have user interfaces for information processing and end-user information manipulation. This focus on representations has not only allowed us to make computational abstractions (of the world around us), but it has, over the last few decades, enabled us to place a model of the world in the computer for further processing. For this book, this is a central dimension. While one such representation-driven approach allowed for representational models of the world to be built in the computer, it also introduced a disconnection between "the world" and that "representation of the world." The representation allowed for easy manipulations, calculations, and visualizations; but at the same time it was exactly that—activities allowed for in relation to the representation, not the real thing. As this book will illustrate, that fact can at a first glance be viewed as a tiny aspect, but as we are increasingly interested in designing across the digital and the physical, it turns out that that "tiny aspect" is in fact a central, and even metaphysical, concern. If a separation between "the world" and "the computer" is an unavoidable dimension of doing interaction design, then can we ever arrive at interaction design solutions that truly place computing out there, as part of our everyday world and as an indistinguishable dimension of everyday objects?

Still, if we look back at the last few decades of designing and building interactive systems, it is clear that our design paradigm has rested on such grounds. The representation-driven paradigm has been strong and we have developed representations of information flows, states, and activities—and we have designed interactive computer systems as to work with such representations. With this clean focus on representations, we have established a design profession and developed design methods to guide us in the creation of models of our world that the computer can process and present back to us in the form of user interfaces. This approach of making abstractions of the world around us, rather than modeling computing as a concrete material of our everyday world, has been the backbone of design tradition all the way back to the early days of computing. Along with the vocabulary of "representation," as a notion that addresses the connection between *something observed* with *a model of that thing observed*, we have introduced "metaphors" and "symbols" to make it easier for users to understand how a particular representation relates to what it represents.

In relation to this representation-driven history of human-computer interaction (HCI) and interaction design, the recently acknowledged "material turn"[1] in our field marked a significant shift.[2] Initially, the "material turn" in the field of interaction design looked like nothing more than a shift toward "materials" as a term to talk about physical user interfaces, tangible user interfaces, and most recently the Internet of Things where computing could be integrated in physical objects. However, and far more important in relation to the design tradition underpinning our field, the "material turn" also marked something much more fundamental. It marked both a shift away from our tradition of working with representations in computing and a shift toward ways of making computing part of what is real. In short, the "material turn" was a move away from any separation of "world" and "representation" (of that world), and a move toward a complete integration of computing in our everyday world (including the integration of computing in everyday physical objects). From this viewpoint, the "material turn" closed the world-representation gap, and it brought computing out of its box and into our physical world. As such, and as this book will detail, this shift dissolved the traditional separation between the real and the computer, between the world and a representation (of that world). By closing this gap, the "material turn" opened up a new design space for interaction design. In this book I will explore this design space under the label of "material-centered interaction design."

From a historical viewpoint, some significant markers of this material turn include how interaction design switched the focus from *metaphors* to *materials* as a starting point for doing interaction design, and how interaction design practice shifted *from representations* to *presentations* (manifestations) in terms of design outcomes. While "the material turn" indeed refocused interaction design toward more tangible forms of computing, it also implied a new interaction design paradigm in terms of design process as well as a new chapter in the history of interaction design philosophy.

Most recently, "the material turn" has marked its place in the history of interaction design through an expanding interest in the development and use of new material-oriented methods and approaches to interaction design—ranging from craft-based approaches to maker cultures to a growing interest in materials, stuff, and repair. In terms of theory development in HCI, "the material turn" has not only manifested itself as a growing interest in articulating how the combination of digital and physical materials in design projects expands the design space for interaction design—ranging from notions of ubiquitous computing to tangible user interfaces—but more profoundly, it has also led to theoretical discussions on how we can understand interaction design in terms of its materiality.[3]

Starting with this notion of "materiality" as the entangled and intertwined relation between material compositions and users/people, I will present in this book a design agenda for interaction design that builds on the principles of

- *Understanding interactivity,*
- *Understanding materials in relation to interaction design,*
- *Designing interactive compositions.*

These principles will be addressed through the introduction of a "material-centered approach" to interaction design. In this book this approach will be presented as a complementary way forward. That is, this proposed material-centered approach to interaction design will be discussed as a design philosophy and interaction design strategy that takes into account the relational aspects of computing, and as such provides an approach that seek to establish bridges across distinct categories of matters (for instance, interaction design that seeks to build solutions across the physical vs. the digital, or the real vs. the virtual).

While "third-wave HCI"[4]—including developments in ubiquitous computing[5] and tangible user interfaces[6]—lead our field toward computing in physical form (by demonstrating how computing power could be embedded in everyday objects, and by showing how physical materials can serve as input/output peripherals to computers), the contemporary growing interest in 3D printing, the Internet of Things (IoT), and growing maker/

DIY cultures suggest that not only can we digitize physical objects or design physical user interfaces to computers, but more fundamentally, if computing is increasingly less about digital interfaces and representations, and increasingly more about objects and materiality,[7] then interaction with and through computers is radically changing as well. In short, the materiality of interaction is radically changing.

In *The Materiality of Interaction* I will place "the material turn" in a historical context. Instead of taking this recent material turn in our field as a point of departure for a richer description of how computing manifests itself at the current moment, this book takes a different angle and addresses the material turn from both a historical perspective and an interactional perspective. In addition to allowing us to see the present, this viewpoint also lets us see how the material turn did in fact change the relation between our material world and computing, and how material dimensions of our world have been brought into human-computer interaction over the history of interaction design.[8]

So another central historical argument in this book is that *materiality has always been a focal concern for interaction design and thus for how we are designing interaction*. But at the same time, I will pinpoint how "the material turn" is now changing this relation between material design and computing, and thus how the material turn has reconfigured the ground for interaction with and through computers. The material turn marked a shift from representation-centered design toward material-centered design, I will argue, but paradoxically, *interaction* has remained our focus whether materiality was mediated through metaphors and representations, or manifested in computational, yet physical, objects.

In the early days of computing, paper-based punch cards were used as input to the computer. These physical cards simultaneously represented data that was already in the world at the same time as they served as the physical means for interacting with the computer. Due to the technology of that time, physical materials (such as paper and wood) were necessary for interacting with the machine. As such, human-computer interaction was very much about material interactions, by necessity rather than by choice. Since then, we have spent the last fifty years trying to make a distinction between the physical and the digital world, and we have invented ways of linking and bridging these two distinct matters.[9] In this endeavor, the notion of *representation* has been key and a central concern for any interaction designer.

Over the last few decades, a wide range of system development methods has been developed with a focus on how to accurately *represent* aspects of our world in information systems—ranging from financial transactions, activities, and information flows to material states, locations, and changes. Under the umbrella of how "*anything that can be measured in the world can also be represented in a computer,*" information systems design became the project of linking our physical world to the world of computers. Representations allowed for this distinction between our physical and digital world. Metaphors, symbols, and linguistics became tools for *bridging* what was considered as two distinct matters—our physical world on the one hand, and the computer on the other. Office computerization, process automation in industry, and modern air traffic control are just a few examples of the strength behind representing states in the world in computer systems—allowing the computer to represent the world, and allowing not only for manipulation of the world via computer systems, but for elaboration on alternative states in the world via simulations.

"Representation" has indeed been one of the central notions for information systems development and for interaction design over the last few decades. The introduction of graphical user interfaces (GUIs) in the mid 1980s pushed the development of representation even further through an increasing interest in graphical representations of digital information, as well as for how the computer could also represent states in the real world.

Most recently, the wide adoption of sensors in cities and vehicles (some people are even talking about "smart cities" and "connected cars") has provided new opportunities for representing states in the world, for collecting "big data" about the world, and for further processing by computers—again presenting new opportunities to represent flows and states in the world with the use of computers. So, although we constantly invent new concepts (like "smart cities" or "big data"), it is still, to a large extent, about arranging computational materials for the production, presentation, and manipulation of representations of activities, information, and data.

Taking place alongside the development of the representation-driven design paradigm was the development of a whole vocabulary for guiding and articulating the design of computerized solutions for representing the world in the computer and for manipulating states and flows in the world via the computer. As this interaction design vocabulary was developed, the notions of *symbols* and *metaphors* filled an important role in translating the mediation between the world and what the computer represented—typically via a graphical user interface. The "mediating role" of information technology, as described in a number of theoretical frameworks from this time, underscored this way of seeing the relation between the computer and the world as two distinct entities, and how these entities were deliberately designed to work in relation to each other.

"The material turn" marked a disruptive change in relation to the representation-driven history of computing and interaction design. System development driven by representation had suggested that good interaction design was about the design of computer systems that (1) could accurately and meaningfully *represent* states in the world, reproduce these states in the computer, and re-represent these states to the user, and (2) could, through *symbols* and *metaphors*, provide ways of manipulating these states with the computer as a tool and a mediating structure. The material turn marked the beginning of a movement away from the focus of interaction design on the virtual, the immaterial, the abstract, or the representational qualities of information, and toward ways of giving the computer, and thus interaction design, physical, substantive, concrete, and material form. It took interaction design away from *representations*, as it manifested a break from *symbols* that illustrated one thing while pointing to a different thing, and a break from *metaphors* as a bridging mediator—between users' understanding of the world and the computer in relation to that world. One of the more common examples is the "desktop" as a metaphor and as a way to both organize and represent information in the form of a user interface.

In dismissing *representations*, *symbols*, and *metaphors* from the interaction design agenda, the material turn did mark a turn *away* from something. In terms of "the materiality of interaction," this turn was fundamental on several levels. While the representation-driven era of interactive systems development upheld a distinction between two entities—between the (material) world on the one hand, and how the computer could be designed to represent this world on the other hand—"the material turn" suggested that any such distinction between entities could no longer be upheld, or defended. Tangible computing suggests that physical materials can be part of a user interface; computational devices, such as smart phones, would make no sense if not also considered in relation to the form factor of the device. And what would the Internet of Things be if computing and things were kept apart as distinct entities?

Over the last few years, computing has changed. It has not radically moved forward; there is no "clean cut" between then, now, and a foreseeable future. Rather, current developments have revisited the early roots of how computing systems were designed across physical, electrical, and digital components. With that as a point of departure, we can now see a wide range of new examples of interactive systems design that build upon such ideas of integrating, rather than separating, matter.

Along with this new mindset, the computer is now reintroduced into the world—the same world it was just recently designed to represent—as yet another design material, and the historical brackets dividing the world into different distinct entities are no longer meaningful, nor possible.

Of course, if interaction design up until this point in history was a project for *"the development of good and meaningful interactive representations"* that allowed for easy, meaningful, and pleasurable manipulation of information, then a shift away from representations toward materials did mark a radical shift away from interaction design as a mediating structure between our physical and digital world and toward a situation where interaction design becomes the project and approach for merging the digital with our material and social world.

However, not only did the material turn reintroduce the computer to the world, making it a part of, as well as integrated and embedded in, the material and social world (under labels such as "ubiquitous computing," "embedded systems," and "ambient and pervasive computing"), but this change of focus also came with some implications for interaction designers. While interaction design for a long time was about this concern for representation and a concern for how we perceive an interface as presented to the user, it also focused on the level of the *application,* i.e., on materials already brought into a composition. The material turn thus implied a change of focus—away from the complete and ready-made applications, away from compositions, and away from a design language which speaks about user interface design in ready-made terms and where the "wholeness of the composition" (that is, the application) can be evaluated and examined as a functional and aesthetic whole (including, for instance, the somewhat established notions of "coherence," "distance," and "balance" in interface design, which all speak to how the interface is visually configured and presented—as a whole). Moving away from these aspects, interaction design has turned toward the "raw materials" of design as a starting point, toward a material understanding of things, and consequently toward material properties, material qualities, and attention to the material properties and details that make the design.

Let me illustrate this argument with a practical example. When Moleskine and Evernote established their collaboration in 2012 with the goal of developing an integrated solution combining Moleskine's physical notebook and Evernotes note-taking app, it was as important to design a stable note-taking digital service as it was to design a physical notebook that the app could read, while not losing the feeling of the traditional Moleskine notebook. OCR-reading of handwriting and the stable cloud-based uploading of notes, for example, were just as important as the "look and feel" of the Moleskine notebook as the canvas for users to interact with this physical-digital composition.

This collaboration between Moleskine and Evernote makes it possible for a user to physically jot down notes in a paper-based notebook, using a smart pen that simultaneously scans and syncs these notes with the Evernote online service, which saves the notes in the cloud for further access and use. A paper notebook, a smart pen, a smart phone, and an online cloud service are deliberately designed here to work in concert to allow people to do computing while interacting via an interface that looks like a traditional pen and what to the human eye looks like a traditional notebook. In this example, the note-taking activity and the notes jotted down in the notebook are not representations of the notes. Nor is "note taking" a metaphor for what the user is doing. On the contrary, "note taking" was the key activity in focus for this design project, and digital and physical materials were deliberately formed and arranged around this activity in order to allow for actual note taking with computer support.

Today we can see this shift from the use of representations and metaphors toward material interaction design happening not only in interaction design research labs, but more broadly in industry as well. The Moleskine-Evernote collaboration is only one recent example in which a traditional product has been enhanced with digital and interactive capabilities.

Across its ongoing digital development, the profession of interaction design is increasingly about combining and merging different materials in design—regardless of whether these materials are "material" or "immaterial" in terms of their appearance or fundamental properties. Interaction design is also increasingly about physical design, fashion, and style—just take the whole trend toward digital jewelry and the wide range of smart watches as an example. Interaction design is no longer restricted to organizing things on the virtual screen, representing information, and enabling those sets of information to be manipulable via a keyboard and a mouse. Far beyond that, interaction design is increasingly about designing a wide range of interactive and computational elements to work completely integrated with other physical materials, objects, and even our bodies.

From the viewpoint of how we integrate different materials, forms, and computational powers in the designs we produce, the materiality of interaction is not only a matter of the material interplay that enables a particular form of interaction, but is equally a matter of how we perceive and make sense of a new design. Again, the introduction of new "smart" watches to the market might serve as an example to illustrate this.

A typical analog clock or watch is round. Accordingly, when exploring the design space for so-called smart watches, a number of tech companies are now using this form factor as a way of helping us understand this class of devices as "watches." The idea is quite simple. If we see an arm-worn object that consists of a bracelet and a round object, we can quite easily interpret that as "a watch." Interaction designers working in this field use the round clock face to help us form our understanding of this new computational object, and hence avoid interpreting the new device as a miniaturized desktop computer rather than as a "smart watch." In short, the material form is key for guiding our understanding of this interactive device. In addition, the form factor borrowed from typical wristwatches guides the design across physical and digital interaction design toward an explicit goal—that is, it should resemble and be interpreted as a watch, but as a watch with something additional (smart functionality). The physical form factor of these computational objects and the tight integration of hardware-software and the physical manifestation of digital functionality (e.g., added functionality beyond setting the time and date with the "crown" of the watch as an input device) are carefully considered to communicate a particular strand of interactive objects. The designs also come in different materials (ranging from stainless steel to aluminum and even 18k gold) with different types of faces and bracelets—not for the purpose of providing a mere case to protect the tiny computer inside the watch, but to give this small object a certain character, a symbolic value, a certain materiality, and accordingly guide our understanding and appreciation of this object. Yet this is only one aspect of what the material turn means for "the materiality of interaction."

This growing interest in combining traditional materials and traditional form factors with interaction design, and the increasing attention to detail which follows from this trend, mark a change of focus and mindset—from design with metaphors as guiding ideas to a sensibility for material qualities, of course. This book goes beyond these obvious aspects of a material turn, however, to consider what material interactions might imply for interaction design, and accordingly how we might interact with and through computers, now and in the near future.

At the moment, the field of industrial design is going through a renewal process as traditional product design meets new opportunities enabled by new interactive technologies, including for instance new sensor technologies, smart materials, and microprocessors. Another example is the growing interest in 3D printers and the use of computers not only for information manipulation but for printing or creating physical "stuff." While the representational era of computing was about observing the world and then building computer systems that could accurately represent the world, the material turn suggested a change in this process, making form-making via the computer central for giving material form and existence to what was first only *in* the computer.

With *The Materiality of Interaction* I have a couple of explicit intentions. First of all, the book sets out to provide a historical overview of how interaction design has moved from a *representation-driven interaction design agenda*, via "the material turn," toward a *material-centered interaction design agenda*. Throughout this history, "materials" have stayed as a central concern for interaction design—but the relation between "materials" and "interaction" has radically shifted. During the representation-driven era of interaction design, *the materiality of interaction* was constituted by materials mediated via representations, and we used symbols and metaphors to ease these translations between the representations created and what they represented (in the world). Through the material turn, we are now rapidly entering a new era of computing and interaction design. In this book, I refer to this new era as the emergence of what I would like to call a *relational practice* focused on "material-centered interaction design." Second, in an attempt to address this process of development, an additional intention with this book is to offer an approach for doing material-centered interaction design, and to offer this as an approach that focuses on

(1) Interaction design with a foundation in material sensitivity,

(2) Interaction design that stays close to the materials at hand, and

(3) Interaction design focused on understanding how digital materials can work in concert with or activate other analog/physical materials.

In short, this book is about how *the materiality of interaction* is radically changing as we move toward material-centered design, and how this new era of *material-centered interaction design* is bringing users and designers in close relation to the materials at hand. This relationship is an ultimate goal in my advocating for craft-based approaches to interaction design; it is also an important goal for an interaction design agenda that has not been able to advance as long as we have upheld categorical distinctions between the digital and the virtual, and between the world and representations (of that world).

Still, it is important to bear in mind that this book does not aim to close these "gaps"; more radically, I suggest that the conceptual construction of these gaps was probably necessary at one point in the history of interaction design in order for us to learn how to work with representations and metaphors, as well as for establishing bridges between "the world" and "the computer," and between the physical and the digital. By introducing these distinctions, it allowed for the creation of computing machinery that could process data (from the world), work with representations (of that world), and present output (to that world). By introducing categorical concerns, we fuel the development of the computational machine, and at the time this was probably a necessity for the development of the computer. However, and beyond that historical parenthesis, this book suggests that any such categorical concerns are now becoming increasingly irrelevant, as "computing" and "the world" are no longer two distinct entities easily framed and distinguished from one another—neither conceptually nor practically. On the contrary, we need to think compositionally about matters in order for the profession of interaction design to move forward.

In terms of the structure of this book, chapters 1 and 2 review the historical roots of the "material turn" in our field, with a foundation in the traditional representation-centered approach to interaction design (chapter 1), and what the material turn implied (chapter 2). In shifting perspectives from representations to materials, chapter 3 details what that shift implies for how we should understand one of the most central notions in interaction design—that is, the notion of interaction. With an informed understanding of how material and interaction can be interrelated, chapter 4 presents material-centered interaction design as complementary to the existing approaches to interaction design. In doing so, this chapter provides the necessary grounds for discussing interaction design across substrates (chapter 5) and interactive compositions (chapter 6). In chapter 6, I also return to the notion of interaction as first elaborated on in chapter 3—both to revisit this notion through a material-centered lens and to theorize interaction in relation to the overarching concept of this book—the *materiality of interaction*. With both the practical and theoretical pieces in place, chapter 7 outlines the broader implications and opportunities for a material-centered approach to interaction design, while the closing chapter—chapter 8—looks around the corner for how to move forward.

1 Representations
HCI and Its History of Representation-Driven Interaction Design

Computing—A material concern?

Computing, and human interaction with these computational machines, has commonly been thought of as an abstract activity where we manipulate digital objects on glossy screens. Although working with the computer is hands-on, computing is still abstract to the extent that the work is about arranging and rearranging "painted bits,"[1] clicking on virtual "buttons," processing symbols, working with representations, and even storing and accessing data in "the cloud." How can all of this be even remotely related to materials? Does it really make sense to talk about *the materiality of interaction* in the context of human-computer interaction (HCI) and interaction design? This seems especially counterintuitive at a time in the history of HCI when we are increasingly exploring expressions of so called "non-skeuomorphic design" (an interaction design paradigm which advocates for interaction design free from any material heritage or associations).[2] In fact, isn't the fundamental idea behind non-skeuomorphic design an interaction design paradigm that seems to stand in direct contrast to any material-centered approach to interaction design? Shouldn't the contemporary trend instead be described in terms of a movement away from materials, and toward interaction design that explores ways of designing for the digital, the abstract, the virtual, the representational, and the immaterial? In short, is *materiality* really a concern for interaction design? And accordingly, is it meaningful to talk about *the materiality of interaction*?[3]

In this chapter, we will delve into these essential questions for interaction design. We will do so by revisiting some examples from the early days of computing; and through one such historical perspective, I will illustrate how computing in its early days was enabled through material configurations. Since then, we have tried to uphold various distinctions between the material and the immaterial, between the physical and the digital, between the virtual and the real (most recently in the debate about skeuomorphic versus non-skeuomorphic design), and so I will show that computing has always been, and will continue to be, a material concern. As such, this book addresses an apparent paradox in our field: computing and interaction with computing seem to be concerned with the immaterial, and yet across the history of computing, and across the full range of contemporary interaction design—from cloud-based online services to the Internet of Things (IoT)—computing is recurrently manifested in material form, and interaction with computers are increasingly physical—from interaction with embedded sensors in our homes and cars to urban computing and smart cities, all the way to the simple small-scale computing involved in digitalized paper notebooks.

The aim of this book is to unpack this seeming paradox of how we think about our interaction with and through computers. Although computing seems to be all about the abstract, the representational, and the *immaterial*, it is through particular material configurations and through the careful arrangement of physical and digital materials that computing is enabled.

In an attempt to unpack this relation between computing and materials, I will take as a point of departure the history of HCI and then slowly move toward a position where we leave behind any distinctions between the physical and the virtual, the digital and the real, and the material and the immaterial. As I will illustrate, such distinctions are not helpful for moving forward. Instead, and as an alternative to making such categorical distinctions, I will present in this book a compositional approach to interaction design, where *material-centered interaction design* will be not an exception in relation to interaction design,[4] but a core perspective to consider in just about any interaction design project.[5]

In this first chapter, I would like to go back to the early days of computing. In searching for material aspects of computing, it does not really matter if we go back to the paper-based punch card computers of the 1970s or if we revisit the first transistor-based computers of the early 1950s.[6] From the viewpoint of this book, what stands out quite clearly is that in those early days of computing, the computer itself was not just a machine deliberately designed to do symbol processing; its configuration was also heavily

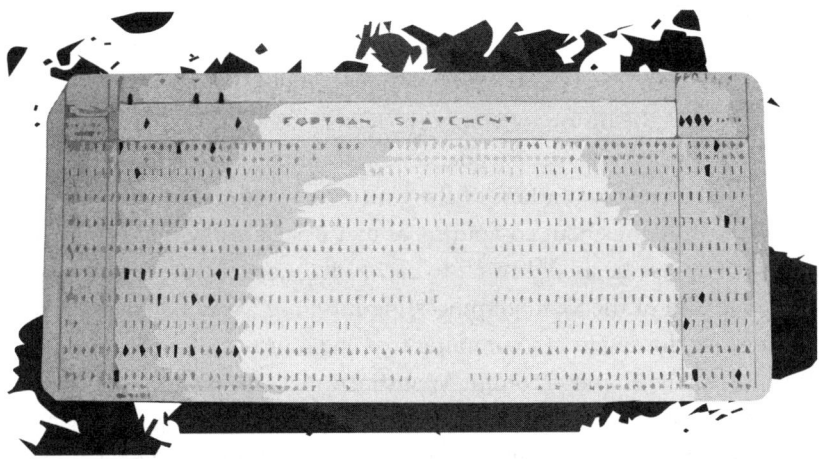

Figure 1.1
A paper-based punch card is an early example of material interaction with data, and of the use of physical materials for computing.

intertwined with our physical world. Computing was a material concern—both in terms of input/output peripherals and in terms of its design and technical implementation.[7]

Right from the start, in the earliest days of computing, we made a number of assumptions about the nature of computing and computing machinery. In fact, as a field comprising computer science, engineering, and design, we assumed that computers should be framed and designed as *symbol-processing machines*—that is, as machinery particularly designed to process and manipulate data sets. Further, we assumed that these symbols *symbolized something*, and that *representation was key* for linking data about that something to a particular symbol for further processing by this computing machinery.

These simple and yet fundamental assumptions about computing became the building blocks for the further development of computing and interaction design for half a decade. If we think about these assumptions and view the history of computing as a project of building the representational symbol-processing machine, something quite fundamental surfaces as an obvious idea for interaction design. Ever since the early days of computing, we have subscribed to a *representation-driven interaction design agenda*; accordingly, interaction design has been about the design of computer programs capable of processing data, about the design of meaningful representations so as to present data as information, and about the means of interacting with these symbol-processing machines.

To say that the history of computing can be described in terms of a representation-driven design agenda, and accordingly a design agenda with an interest in representations of the world (as a slightly different perspective than thinking about computing as manifested in the physical world), might of course be a bold thing to say, so let me elaborate a little bit on this. The idea of the "symbol-processing machine"[8] was a deliberate guiding idea for the design of a computer that could process symbols, combinations of symbols, and accordingly, abstractions. This guiding design idea enabled the design of a computer that could deal with representations of the world, described via well-defined symbols, and so this model allowed for a separation between the world on the one hand, and representations of this world on the other hand.

A physical symbol-processing machine—A paradox?

Somehow the idea of a physical symbol-processing machine—in other words, a computer—almost represented a paradox in the early days of computing. While being implemented as a physical object, the computer was deliberately designed to process abstractions. It was situated in the material world, but it was aimed at processing representations of the world rather than producing physical goods (as any other machinery would). The design of the computer—as a symbol-processing machine—was the starting point for a line of developments which upheld a distinction between the real world and representations of this world—while simultaneously manifesting machinery that was heavily dependent on physical matter.

Paradoxically, while this machine was deliberately designed as a general-purpose computing machine meant to deal not with a particular instantiation of our world but with symbol systems and representations of our world, its essential design was still heavily dependent on physical matter. This paradox is key for understanding the history of computing, and is a key argument throughout this book. From the IoT to cloud computing, materiality is essential for understanding computing in context and for understanding interaction with computing machinery, even today.

Keeping things apart!—A history of representations, symbols, and metaphors

For computing, the notions of representations, symbols, and metaphors are essential. Just try to imagine a computer without these fundamental elements. I would say that is virtually impossible. Representations enable us to create an image of some selected aspects of our world, and the presentational format selected (for instance, a diagram which illustrates some data about the world) in turn has impact and effect on how we interpret, and ultimately understand, not only the representation but the world (although seen through a representation). When we design interactive artifacts, screen-based user interfaces, and so on, we typically rely heavily on different representations. For instance, when we follow and apply the WYSIWYG (what you see is what you get) paradigm of interaction design, the underlying assumption is that there is a 1:1 correspondence between "the world" and "the presentation" of that same world—in other words, "what you see" equals "what you get."

While representations, symbols, and metaphors enabled computing in the form of abstractions, data modeling, processing, and manipulation, these relational distinctions between "the world" and "the model" (of the world) simultaneously led us onto a track of vocabulary development that has since been reflected in just about any new application area of computing. For instance, when we as a field of research got interested in VR (virtual reality), we initiated many discussions of whether virtual reality was different from the "real" reality, and if so, then how. While such discussions can be interesting to have, it is even more interesting to notice how this separation between the virtual and the real reappears in the context of VR. In similar ways, we have introduced distinctions not only between the virtual and the real, but between the physical and the digital.

Of course, our field has also made several attempts to bridge these categorical matters.[9] For instance, the notions of *mixed reality*, *augmented reality*, and *embedded systems* all point to bridges across this fundamental distinction between the digital and the physical, a distinction that has existed since the early days of computing.

However, the problem with these notions is that although the attempt is to unify and work across any categories introduced, the very notions themselves manifest separation rather than integration. Take the notion of "augmented reality" as an example. This notion makes an implicit assumption that there is a (real) reality, and that computing can be "added" to this reality, so as to "augment" it. Clearly this notion contains a separation between reality and computing. A similar problematic unfolds in the notions of "mixed reality" and "embedded systems." The separation between terms is upheld even though these notions have been developed explicitly to work as "bridges across" instead of keeping things apart. Even our vocabulary has been designed and developed to distinguish between things rather than working as guiding frameworks for compositionally unifying matters.

So how do we move forward from this problem? If we look at the history of computing, we will see that we have done more than merely invent new concepts in our attempts to integrate computing with our material world. As the next section will show, we have also been through a type of hands-on journey to establish links between computing and our material world.

Establishing bridges—Linking the physical and the digital world

While the history of computing illustrates the separation made between the physical and the digital world, and while the physical instantiation of the general symbol-processing machine suggests that physical matter has always been a fundamental necessity for computing, a number of attempts are nevertheless currently being made to work across the physical and the digital. These attempts work under a set of different labels, including sensor technologies, embedded systems, pervasive systems, and so on. Even in the early days of computing, not only did the fundamental notions of *data* and *representations* enable a separation between computing and the world, but these notions were dual to the extent that both data and representations also worked as bridges between these two entities.

In fact, *data* is useless if it lacks any relation to the world it is part of. Data is always about the world, and knowledge about what aspects of the world a specific data set covers is key to understanding and interpreting data, and accordingly for turning data into information. The same goes for the notion of *representation*. A representation is always a representation of some aspects of the world. If you know what a representation is supposed to represent, the interpretation is easier than if that essential information is lacking. Clearly it is an advantage to see things as relational compositions.

Representations do not merely manifest a relation between the representation and what it represents. In the world of computers, representations enable us to see what the computer is processing and enable us to have user interfaces in relation to computer program execution. Throughout the history of human-computer interaction, we have typically focused on visual representations,[10] but in HCI we have also explored alternative representations of what the computer processes and ways of engaging with the computer. This ranges from audio-based user interfaces to gestures and even location-based user interfaces where the coordinates of a person also work as input to a computer.[11] Whether the representation is screen-based, audio-based, or location-based, representations work to create a bridge between what the computer does and the outside world; they enable our interaction with computers. Not only has this notion of *representations* been useful for the development of computing, but this fundamental idea of representations has also lead to the development of a language, or vocabulary, for talking about how to turn aspects of the world into a computable format— that is, a form and a material manifestation that we can interact with.

While data and representations enable bridges between the computer and the world, contemporary developments address ways of working across our computational and physical world in a more direct manner. Sensor technologies, the Internet of Things, embedded systems, ubiquitous computing, ambient displays, interactive architecture, and mobile computing are all examples of more hands-on, and more direct, computing in the world, with the world, and through the world.

In between the early notions of *data* and *representations* and today's notions of, for instance, "urban computing" and "smart cities," we have yet another notion that works as a bridge between the world, on the one hand, and the world represented in the computer on the other. This powerful notion is the notion of *metaphors*. Ever since the first attempts to design graphical user interfaces (GUIs), we have followed a design idea of "recognition rather than recall."[12] Instead of requiring users to remember a wide range of commands to interact with a computer via a command line interface, we have designed user interfaces which users can understand based on recognition. Accordingly, if we rely on metaphors in user interface design,[13] we can understand functionality associated with "buttons"[14] through an understanding of how things work in our physical world. So, if we introduce a "button" in a user interface, we assume—based on our understanding of how buttons work in the physical world—that we can click on it, or use it to turn something on or off. In short, the frequent use of metaphors in user interface design has helped us to understand and interact with the computer through our understanding of how things work in the physical world.

Moving from the physical world to the digital world—"Skeuomorphic design" as the design paradigm for grounding interface design in the physical world

While metaphors enabled design of graphical user interfaces, which on a metaphorical level remind us of the physical world, even more hands-on attempts have been made in the history of HCI and interaction design to use material dimensions and characteristics of the world as a basis and point of departure for the design of graphical user interfaces. Such attempts to ground digital design in the physical world as its raw model have recently been labeled *skeuomorphic design*.[15]

Skeuomorphic design is the method of designing something to look as if it is made from real-world materials; skeuomorphic interaction design is thus interaction design deliberately focused on designing user interfaces that, although presented in a digital format, give the impression of being made out of physical materials. For instance, an e-book does not by necessity need to look like an old paper-based book with pages that can be flipped through, turned, and bookmarked with placeholders. Neither does a digital notebook app need to look like a paper-based notebook with a stitched leather cover and linear paper. However, for a long period, skeuomorphic design has ruled graphical interface design; figures 1.2 and 1.3 might serve as illustrative examples of this design paradigm.

The idea of skeuomorphic design enabled ways of designing graphical user interfaces with seemingly physical qualities. While it added a physical flavor to interface design, it is important to remember that this was still in line with a representation-driven interaction design agenda. The iPad notebook was designed to look like a traditional paper-based notebook, for example; it was designed as a representation of a notebook.

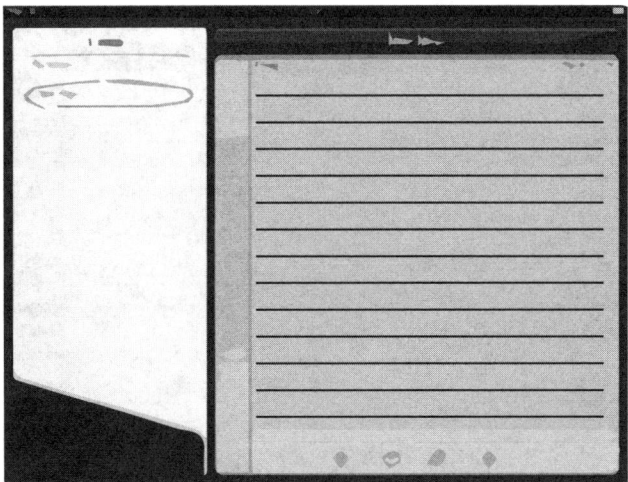

Figure 1.2
The Apple iPad Notes app is an example of skeuomorphic interaction design.

Yet another example of *skeuomorphic design* can be found by looking at how scrolling is implemented on the Apple iPhone and iPad. When you try to scroll beyond the top or bottom of a list, the list bumps back as if there was a physical force at play with it.

A key to creating such illusions of physicality in the graphical user interface is thus to answer design questions such as: How do we represent physical matter, such as paper, leather, stitching, gravity, etc., on a digital screen? How do we create the experience or the illusion of a leather notebook on a digital screen? How do we design the experience of the e-book so that it feels as if we are flipping through an old paper-based book? In short, how do we translate aspects of the physical world into a representation that manifests this bridge between the world and the re-representation in the computer?

Bringing aspects of the world into the computer definitely causes design challenges—not least because we designed this gap or distinction between the world and the computer in the first place. Although in the early days of computing we made a number of assumptions about how the computer could uphold a distinction between the world and the representation, so as to make things computable, we are now witnessing a similar assumption being made about how we might design the digital without

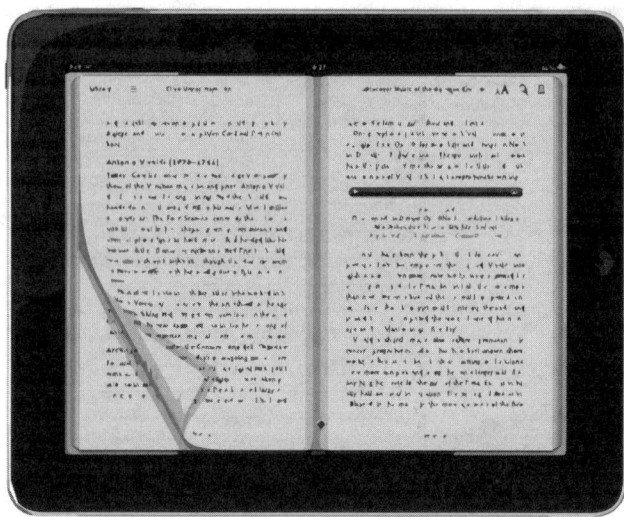

Figure 1.3
The Apple iBook is another example of skeuomorphic interaction design where the user can flip paper-like pages in an e-book.

any traces of the physical world. This alternative is the design idea of "non-skeuomorphic design"—a design idea that we will look at in more detail in chapter 4.

It is interesting to notice how the field of HCI research and the practice surrounding interaction with computers have repeatedly revisited the idea of keeping the physical and the digital apart—as if we did not have any other way of thinking about computing and its role in our society, in our cultures, and in our everyday lives. However, beyond any such categorical distinction between the digital and the real, this book advocates for an alternative paradigm for interaction design. Starting with the notion of "materiality" as the entangled and intertwined relation between material compositions and users/people, I will present a design agenda for interaction design that builds on the principles of

• *Understanding interactivity,*
• *Understanding materials in relation to interactivity,*
• *Designing interactive compositions.*

In the coming chapters of this book, I will address this agenda through the introduction of a "material-centered approach" to interaction design as an alternative way forward. In short, this proposed material-centered approach will be presented as a design strategy that takes into account the relational aspects of computing instead of aiming to bridge distinct categories of matters (e.g., physical versus digital, or real versus virtual). As I will argue, not only do we need a new method and approach for doing this, but we also need a new way of seeing and talking. In other words, we need to see the design of computing through a compositional lens, and develop a compositional language and vocabulary[16] in support of compositional interaction design thinking.

But before jumping to conclusions, the next chapter will talk about "the material turn" in our field and what that implied in terms of an increasing need for thinking compositionally across different types of matter.

2 The Material Turn
On the Material Turn in Interaction Design

Whether we think of interaction design as a design tradition aimed at giving form to interaction with and through computational objects or as being about user interface design, it is hard to escape the fact that the user interface to a large extent defines the scene and the form of the interaction. Without adopting a fully deterministic perspective, it is nevertheless a fact that if the user interface is screen-based and graphical and the input modality is mouse-based, then it is likely that the form of that interaction—that is, what the turn-taking looks like and what is demanded by the user—is very similar to other screen-based interfaces with similar input devices. However, the design space for the form of interaction is growing quickly. While command-based interfaces and text-based interfaces defined almost the whole design space in the 1970s, developments since then, including novel ways of bringing sensors, actuators, and smart materials to the user interface, have certainly opened up a broader design space for interaction design.

But it is not only the range of materials that has been extended over the last few decades; we have also moved through a number of form paradigms for interaction design. By "form paradigms," I mean *overarching ideas for how computational resources can be brought together, arranged, and integrated to support, define, and communicate a particular materiality of interaction.* This has been important work because it is through these form paradigms, manifested in classes of interactive systems that share the same type of materiality, that users can make sense of, approach, and use different forms of computer systems. When I say "classes of interactive systems," then, I refer to *broad sets of interactive systems that build on the same type of materiality of interaction.* For instance, command-based user interfaces are one class where interaction with the system is text-, screen-, and typing-based; and the graphical user interface that follows the idea of WYSIWYG (what you see is what you get) is another class, in which interaction with the system builds upon the principles of direct manipulation, but meaning is presented in the form of visual menus and icons rather than being text-based, and the materiality relies on metaphors rather than linguistic grounds.

With this as a foundation, this chapter will reflect on how we have moved from the early days of command-based user interfaces, via the use of metaphors in the design of graphical user interfaces (GUIs), toward ways of interacting with the computer via tangible user interfaces (TUIs). Later, I will describe how this movement toward TUIs was a first step away from building user interfaces based on representations and metaphors, and a first step toward material interactions.

The use of metaphors in interaction design is part of a well-established design tradition.[1] Across generations of operating systems and applications we have continued to rely on the use of metaphors in the design of easy-to-use interfaces for interacting with computers. The "trash can," the "desktop" metaphor, "folders," and the Save icon in the form of the classic "floppy disk" pictogram from the mid 1990s are just a few examples. Such metaphors have enabled us to design graphical user interfaces that the user can understand through processes of association and by recognition rather than recall.[2]

According to *Webster's*, a "metaphor" is

• *"A word or phrase for one thing that is used to refer to another thing in order to show or suggest that they are similar";*
• *"An object, activity, or idea that is used as a symbol of something else."*

For instance, in the context of interaction design, if there is a graphical representation of a trash can on the computer screen, and if this representation of a trash can is implemented so that the user can "drag" virtual "documents" to the trash can and then "empty" it, thus erasing the documents, then the user, knowing how a physical trash can works, can assume that this digital representation will work in a similar way.

As this example illustrates, metaphors can help guide a user in how to understand and use a particular function, a complete user interface, or even a digital service. For an example of the latter, just think of "the cloud." This most recently introduced metaphor suggests that data should no longer be stored in one particular machine, but rather in a more widespread network of computers. The "cloud" metaphor suggest that the files are stored somewhere (in this cloud), and that the particular location is irrelevant from the user's perspective. Clearly, the metaphors we have selected have had an impact on how we understand different user interfaces and digital services. As I will illustrate, this also holds for how we have changed the materiality of interaction as we have moved from the development of one class of interactive systems to the next.

As we moved from the early command-based interfaces of the 1970s to graphical user interfaces introduced in the early 1980s, we also moved from "recall" to "recognition" in terms of knowing how to operate the computer.[3] When interacting with a computer via a command-based interface, the user needs to know and remember which commands the computer will understand and what each command will do, and how different commands can be combined for more advanced interaction with the machine.

From an interactional viewpoint, to rely on "recall" is quite different from using graphical user interfaces (like Windows, Mac OS, Android, etc.) that rely on "recognition." For instance, if the user wants to save a document in the recognition-based interface, there is no need to remember the command for doing so. However, the user needs to recognize something in the user interface that resembles a function for saving the work. This is where metaphors come in. By designing with tight coupling between things in the physical world and representations of such objects on the computer, and by relying on metaphors to ease the transition between the user's understanding of real-world objects and how to operate something similar in the computer, we have been quite successful over the last thirty years in building easy-to-use interfaces for interacting with computers.

This has been guided by the development of design guidelines for graphical user interfaces and of methods, techniques, and technologies for usability testing, including eye-tracking technologies, usability labs, etc. But the basic idea underpinning the design of graphical user interfaces has remained the same, i.e., to work with *recognition* rather than *recall* as a fundamental design principle, and to rely on metaphors in order to design understandable and "intuitive" user interfaces.

During the 1980s and '90s, the interest in usability testing and the development of usability methods was huge in the field of human-computer interaction,[4] in both research and industry. Usability testing was deployed as an integrated part of any development project, and usability labs and integrated hardware-software solutions were developed for usability testing.[5]

Of course, this interest was motivated by the transition from command-based interfaces to graphical user interfaces, but it was also fueled by the fact that the computer had become an everyday item for just about anyone to use through the introduction of the personal computer, or PC, in the mid 1980s. From being a professional tool that required professional training in order to operate it, the computer became an everyday object that ultimately should be self-explanatory in terms of how to operate it—or at least this became a design goal. In this endeavor usability testing became one way of measuring just how easy or hard it was for a user to understand and use the computer for a particular task.

Alongside usability there was of course also a growing interest in "learn-ability" and an interest in finding ways to make it easier for users to move from a novice level of interaction to an expert level.[6] The use of Photoshop has perhaps been the most common example of the full spectrum from novice users to experts.[7]

Since these computer programs so heavily relied on metaphors, the understanding of these computer programs was appropriately based on "understanding through reference." For instance, if the user had experience with discarding paper into a traditional trash can, then a digital representation of one could be fairly easy to understand. The problem, however, was this detour via the physical world in order to understand the digital world. Metaphors as such helped to establish a divide between our physical and digital world.

From metaphors to materials

One of the main implications of the "material turn" in our field was a turn away from metaphors as a basis for interaction design, and a turn toward interaction directly with and via physical materials.[8] In this turn, direct interaction with the materials at hand was preferred over interpretation of one virtual thing via a different physical thing. As such, the material turn also implied a shift away from representations and metaphor-centered interaction design and toward ways of interacting with physical things and objects.

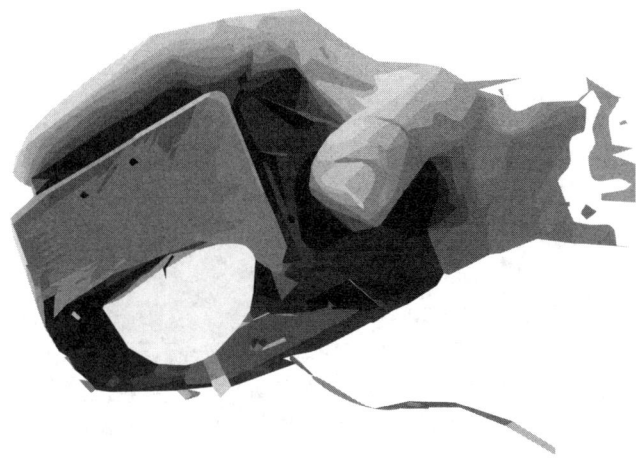

Figure 2.1
An illustration of the original computer mouse developed by Douglas Engelbart in 1964.

It is, of course, hard for a computer to read and operate physical materials right away, so in order to make that happen, these materials have to be "activated" with the help of sensors and actuators—including the full range of technologies from simple RFID tags, NFC (near field communication) protocols, Bluetooth, and WiFi to more advanced positioning systems, etc.—in order for a CPU (central processing unit) to bring these physical materials into the processes of computing. The first computer mouse, designed by Douglas Engelbart in 1964 from a piece of wood, can serve as a paradigmatic example: a simple piece of wood and its movement on a flat surface served as a new input modality for interacting with a computer (see figure 2.1).

The moving around of this mouse was not implemented in any metaphorical way. Interaction via the mouse was actually done by physically moving the mouse across a flat surface—just as we do with any modern computer mouse. In this sense, interaction with the computer occurred directly, with the materials at hand, not via metaphors.

Another illustrative example here might be the Apple I computer. The first computer developed by Apple, the Apple I was designed and hand-built by Steve Wozniak. It was released by the Apple Computer Company (now Apple Inc.) in 1976 (see figure 2.2). The whole computer was made

Figure 2.2
The Apple I computer of 1976.

of a piece of wood, and interaction with this computer was via a typical keyboard.

From the viewpoint of the material turn in our field, however, we need to think critically about whether these two examples actually count as material interaction. Of course, both the mouse and the Apple I computer were implemented in a natural material (wood). However, this particular material, and this particular choice of wood, had no impact on the user's interaction with the computer. Although both the computer mouse designed by Douglas Engelbart and the Apple I computer were made of pieces of wood, there was no real interaction with the wood per se. The wood worked more like a container for the electronics inside it, and the interaction was still to a large extent screen-based. In fact, the container for the mouse and this computer could have been plastic or made of just about any material; the choice of material would not have had any effect on the interaction.

If we compare Engelbart's mouse and the Apple I computer to the Moleskine Evernote notebook described in the introduction, we should acknowledge a couple of fundamental differences from the viewpoint of material interaction design. First of all, the interaction in the case of the Moleskine Evernote notebook is actually with and via the physical notebook, rather than by using the notebook in order to interact with some information on a screen. By scribbling on the notebook, the user is interacting with the

Figure 2.3
The Moleskine Evernote smart notebook.

system. Accordingly, the interaction happens through turn-takings with the notebook itself. Here, "note taking" is not just an activity but also the modality for interaction—the notebook is the user interface, and the output is the notes that were written.

Furthermore, the Moleskine Evernote notebook supports tagging of different pages and sections, and the Evernote mobile app is designed to recognize handwriting. According to Moleskine's website, the Evernote smart notebook "features unique 'Evernote ruled' . . . page styles with dotted lines designed to ensure a clean image when digitally capturing your notebook." And when a page is captured with the Evernote app, "the Smart Sticker icons become searchable, digital tags that allow users to organize notes and to sync the digital and analog notes" (see figure 2.3). So, in a sense, the notebook itself is a computational canvas for note taking, or put slightly differently, the materiality of interaction here is the classical notebook, although reimagined and reactivated as a user interface.

Finally, and maybe most importantly from the viewpoint of how this interaction is experienced the note-taking activity with this notebook does not "remind of" or reassembles traditional paper-based note taking, no such loop of *interpretations and understanding via associations* is necessary—note taking is the activity, not a metaphor for it, and accordingly the activity is not just something similar to note taking—instead, *note taking* is the real interaction here.

Returning to the example of the computer mouse designed by Douglas Engelbart, at a first glance we can notice something similar to the note-taking example. The user interacts in a very direct way with the computer mouse. He or she moves the mouse around, not metaphorically but physically, so in that sense using the mouse is a material interaction. However, a main difference between the notebook example and both the computer mouse and the Apple I computer examples is that in the note-taking example, the materials matter in how they come into play in the interaction. For the computer mouse and the Apple I computer, the wood was a necessity for the design, as these two designs needed physical containers for their hardware, but it did not really matter that the material was wood; plastic or any other material would have worked just as well. The note-taking example is distinctly different, because part of the interaction design is made to actually enable computer-supported note taking on real paper. Accordingly, the paper had to be specifically selected and designed in order not to lose the user experience dimension of traditional note taking. With this as an illustrative example, I would say that what marked the material turn in our field was a turn toward a design paradigm where we started to include certain physical materials in processes of digital interaction design because these materials (like traditional paper) had particular physical properties which could be used in the design of a particular thread of interaction.

Reflecting again on the early computer mouse in relation to the material turn, I would say that, on the one hand, the mouse represents a very direct and physical way of interacting with a computer. On the other hand, the mouse also introduces some aspects of representation, which is the whole idea behind the computer mouse in the first place. While the physical computer mouse is all about physical interaction, the virtual mouse pointer on the screen is at the same time all about representation. The virtual mouse pointer represents and mirrors the physical movement of the mouse across a flat surface, and the virtual mouse pointer gives access to the further manipulation of other representations on the screen.

So, while it can be said that the user is interacting directly with the computer mouse, he or she is only indirectly interacting with the information on the screen. Here, manipulation of information on the screen is mediated via the virtual mouse pointer.

To get a historical understanding of what the material turn is all about, I would like to look back to Shneiderman's early work where he tried to define what "direct manipulation" is in the context of interaction with computers. The notion of direct manipulation is, to a large extent, related to a strand of interaction design thinking that wants to put the user in direct contact with the (computational) materials at hand.

As described by Hutchins, Hollan, and Norman (1985),[9] the term "direct manipulation" was coined by Shneiderman (1982; 1983) to refer to systems with the following properties:

(1)   Continuous representation of the object of interest,

(2)   Physical actions or labeled button presses instead of complex syntax,

(3)   Rapid incremental reversible operations whose impact on the object of interest is immediately visible (Shneiderman 1982, 251).

Here we can see some early signs of the historical movement toward the material turn. Physical actions, and the directness both in terms of material engagement and feedback, point toward a direct, rather than a mediated or interpretive, model of interaction.

In one sense, the computer mouse fulfills the definition of direct manipulation as provided by Shneiderman (1982). The virtual mouse pointer continuously represents the movements of the physical mouse. Physical action in the form of physical movement or clicks on the mouse is the way of operating the computer instead of using complex syntax, and rapid operations are immediately visible on the screen.

However, from the viewpoint of material interactions, that is as close to direct manipulation as the mouse gets. The mouse is still a means, a mediator, for moving the virtual mouse pointer.

So how can we move beyond this mediating role of technology? How can we reach beyond any representation-driven approach to interaction design? Can material interaction ultimately be a way for users to directly interact with computational materials without any mediators in between the user and the computer? And, as a final question for the moment: Are there more dynamic examples than the Moleskine Evernote notebook when searching for an alternative paradigm to interaction design built around representations, symbol manipulation, and the use of metaphors?

These are just a few questions that can be raised when trying to distinguish representation-driven interaction design from material interaction design. While these questions might be hard to answer right away, this book, through examples and through the development of a methodology, will present an account of material interaction design and models for doing it as a distinct design paradigm.

For instance, and to add some more examples right away, let us take a look at two recent research projects, called inFORM[10] and TRANSFORM,[11] both led by Prof. Hiroshi Ishii at the MIT Media Lab. At this moment in the history of human-computer interaction, the inFORM project might be one of the most obvious examples of material interaction and an illustration of what the future could hold for material interaction design.

The inFORM project has a tabletop tangible user interface that is a shape-shifting physical display which the user can operate by physically touching the surface of the table, and through such means can manipulate tangible bits with his or her hands (see figure 2.4). The fundamental philosophy behind this project is Ishii's vision of "radical atoms."[12] Beyond the design of tangible user interfaces which establish bridges between atoms and bits,[13] the idea of "radical atoms" is to reactivate physical materials (atoms) with computational power.

So what can this example tell us about the material turn in interaction design? Well, first of all, it illustrates a physical interface that is also simultaneously the output of the interaction. This table is not a new kind of input peripheral to a computer monitor or any other form of graphical display (as the computer mouse was when it was introduced in the mid 1960s). Beyond that, the user is in direct contact with the materials at hand and can directly—and with his or her bare hands—work with this physical presentation of the computer.

Furthermore, the inFORM project illustrates how the material turn moves interaction away from mediating tools and objects (including the keyboard, mouse, and other such means) and toward more direct forms of interaction, and in that sense also toward more embodied ways of interacting with a computer. This project illustrates that computing can come in many different forms—including many different physical forms—and as such, it illustrates that the typical desktop computer "box" (which I imagine many of us think of when thinking of a typical computer) might no longer work as the only form factor for how computing can manifest itself in the world. In short, the inFORM project suggests that computing can take any physical form, that interaction with computational objects can be

a.

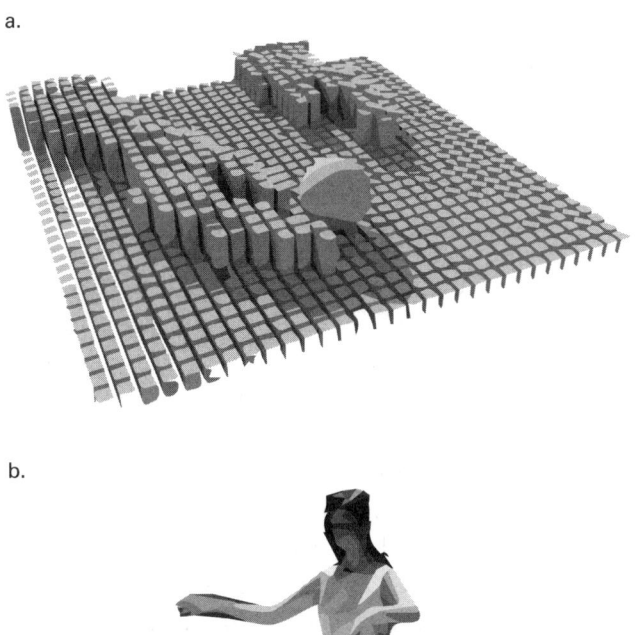

b.

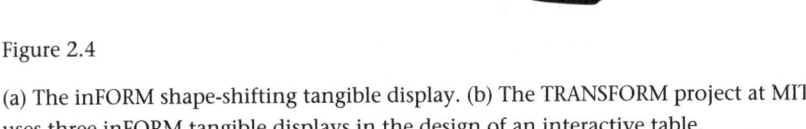

Figure 2.4

(a) The inFORM shape-shifting tangible display. (b) The TRANSFORM project at MIT uses three inFORM tangible displays in the design of an interactive table.

direct, and that computational manifestations can work as a complementary design paradigm to the more traditional representation-based design paradigm.

From representations to manifestations

Speaking of "the box" and alternative ways for the computer to manifest itself in the form of a physical object in the world, this challenge of reimagining the traditional rectangular box is sort of obvious right now, when people are amazed by round clock faces on so-called smart watches. Can it be, however, that this amazement is grounded in a preset understanding of computing as always coming in a rectangular box shape? And thus, that the round clock face breaks that understanding of the smart watch as a computer, although a miniaturized one?

At the same time we should bear in mind that monitors have not always been rectangular. In fact, if we go back to the early days of television, we can find many examples of round monitors, of which the vintage Zenith is just one (see figure 2.5).

The point is that as we move from a representation-driven interaction design agenda to material-centered interaction design, our design space is not automatically limited to manifesting computing in the form of a "box." Instead, computing can nowadays take almost any form, and the range of possible manifestations is almost endless. But our previous experiences of a particular technology influence our imagination of what forms it can take.

So, while discussing manifestations of computing in physical form and this movement from representation to manifestation, this might be a good time to ask a fundamental question. In short, what is a "manifestation"? (And is that the vocabulary we should use when talking about how computing presents itself in physical form?) According to Oxford Dictionaries, a manifestation is "an event, action, or object that clearly shows or embodies something abstract or theoretical."[14] This definition works quite well for discussing the turn from computing in the form of abstract manipulation of symbols and the use of representations to thinking about how computing can be embodied in physical forms and in physical objects—i.e., to move computing from the world of the abstract and the representational to the world of the concrete and presentational.[15]

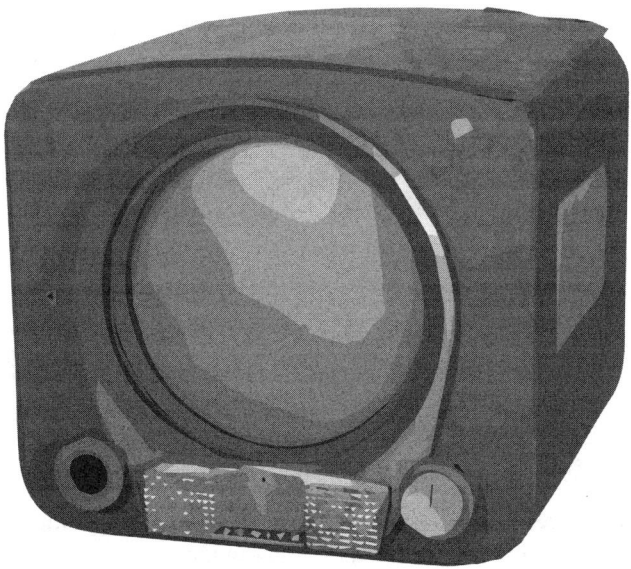

Figure 2.5
The vintage Zenith round-screen tabletop TV.

So, if manifestation from this perspective is about making computing more concrete, it is simultaneously about collapsing "the virtual" and "the real." From the viewpoint of material interaction design, this is an important ontological matter. If computing is less and less about representing the world in the computer, then it is becoming more about this imagined gap between the physical and the digital and between the real and the virtual.

Integrating the computer in the world

A modern car might serve as a good example of how computing is increasingly becoming integrated in our everyday world. Who thinks about interacting with computers while driving a modern car? Still, during a normal drive, we are in fact interacting with multiple computers simultaneously. Besides active brakes, we have automatic obstacle detection, AC, lane-keeping radar, cruise control, and so on.

Through the material turn, the computer has manifested itself in many different forms, and in many cases it might not come in the form of a stand-alone object but rather be completely integrated into an everyday object, a vehicle, or even a building.[16] Computing brings with it the potential to

Figure 2.6
The Mogees device, together with a mobile app, enables its user to turn any object or surface into an instrument for playing electronic music.

activate or reactivate traditional materials and objects and bring them into computational compositions. To illustrate this, we can take a look at the Mogees system.

The London-based computer music researcher Bruno Zamborlin invented Mogees. The system "combines a contact microphone with an app that analyses the vibration of objects and turns that data into musical sounds" (Turk 2014). By attaching this vibration sensor to any everyday object, the object can be turned into a musical instrument (see figure 2.6).

This example illustrates not only how any everyday physical (and analog) object can be activated to serve as a musical instrument; it also illustrates that any everyday object can work as input to a computer, and thus any everyday object can be part of interaction with a computer. Beyond the traditional mouse and keyboard interface, a pan or even a box of nails can serve just as well as an appropriate input device and interaction modality.

The example of the Mogees system also serves to orient us toward the qualities of objects and the properties of materials. For instance, if we want to create a maracas with this technology, we should probably choose an object or a texture that can generate similar vibrations so that the Mogees system can detect that as input for further sound processing.

Figure 2.7
A 3D-printed electric guitar.

Alternatively, we might want to construct material containers that give the technology a well-known shape. Here the trend toward 3D printed guitars might serve as one illustrative example (see figure 2.7), and yet another example is the outer-space-inspired 3D-printed violin by Eric Goldemberg at Florida International University and his MONAD Studio partner Veronica Zalcberg.[17]

Beyond any such skeuomorphic designs, in which the designs we generate have their origin, or any clear heritage in objects of the physical world, this tight integration of computing with our physical world enables us to start exploring radically new designs. For instance, still within the context of musical instruments, we can explore a design space of computationally enabled instruments that we have never before seen and sounds we have never before experienced.

As the computer gets integrated into our everyday world, not only finding its way into our everyday activities, habits, and cultures but more literally "taking place" in new ways, it makes sense to talk about new manifestations of computing in our everyday world. A *computational manifestation* is the physical form or expression computing takes and how it places itself in our material world. Accordingly, a computational manifestation is the particular physical material at play to make computing a physical reality. Interaction designers focused on placing computing out here in our everyday world must think not only about what information the computer will process, but also about how this processing should be presented in material form. Material selection is an important part of the design process.

Furthermore, it is not sufficient to consider only different materials (like wood, steel, concrete, or glass); designers must also think about material properties. For instance, is it a bendable material? Or squeezable? Maybe a sound-absorbing material? Does it have any other necessary properties that can be brought into composition with other digital materials in the creation of a particular user experience? In short, when doing material interaction design—materials matter!

As interaction design enters areas that more traditionally are sectors of industrial design (such as product design or vehicle design) or other areas of design such as clothing, fashion, and fine arts, it has expanded from being a design discipline heavily focused on the immaterial, the abstract, and the symbolic toward also simultaneously embracing the physical, the material, and the real. However, it has not given up its original orientation; such a radical shift would eliminate what is unique about interaction design. Instead, the combination of these two traditions is what makes material interaction design so interesting. Material interaction design sits right in between traditional interaction design and traditional industrial design. It is in the blending of the material with the immaterial, in the combination of abstract information processing with ways of physically manifesting computing, that material interaction design has its unique position. And already today some interaction designers have started to explore this unique design space that sits at the intersection of abstract and concrete forms of computing. Some people refer to this space as a new computational aesthetics,[18] while others have described it in terms of a new area of form giving.[19]

One such example of the combination of interaction design, fashion, and materials in the form of fabrics and biomaterials is the bioLogic project at the MIT Media Lab.[20] In this project, a group of researchers led by Hiroshi Ishii has created some wearable computational textiles, such as a sports jacket that "breathes" in a novel way. In fact, this jacket actually "breathes" via bacteria-powered 3D-printed biofabric that opens and contracts when you sweat while you exercise (see figure 2.8).

Figure 2.8
The bioLogic project by the MIT Media Lab. A bacteria-powered 3D-printed biofabric opens and contracts as you sweat.

While the scenario for using this jacket is interesting, the bioLogic project is also interesting from a material interaction design perspective. Natto cells are used as nanoactuators in the design of this shape-changing interface. Through this manifestation of new materials in the form of a jacket, this project illustrates again that computing and new materials can come in many different (physical) forms and that (physical) materials can be brought into composition with new smart materials and computational power.

In summary, material interaction design opens up a design space where we can explore new forms, functionalities, and appearances of computing across physical, digital, and even immaterial materials (e.g., distance, position, timers, radio).[21] As such, material interaction design is very much about bringing different materials and elements together into formal relations.[22]

As Mark Weiser has said, "the most profound technologies are those that disappear. They weave themselves into the fabric of everyday life until they are indistinguishable from it" (Weiser 1991). As this becomes ever more the case, we need to give up the idea that computing always comes in the form of a rectangular box. In fact, material interaction design is an approach that suggests that "there is no box."[23] Nowadays computing is everywhere and can present itself in almost any material or immaterial form.

In this chapter, I have discussed not only what the material turn means in relation to the shift from a *representation-centered design agenda* to a *material-centered design agenda*; I have also discussed the notion of "direct manipulation" and how the material turn moves beyond mediating tools for direct manipulation and toward even more direct ways of interacting with computers.

In approaching the question of how "close" we are to the materials we interact with, we need to ask yet one more fundamental question: Can we really distinguish interaction from the things we interact with? Or, formulated slightly differently: Is interaction distinguishable from the people and the objects that are part of the interaction? Or, in terms of "closeness": Can we really speak about interaction in terms of "distance"? Or is it perhaps better to discuss "the materiality of interaction" as a unit of analysis and as an object for interaction design where no separation between different elements (human, technology, modes of interaction, objects, etc.) is made in the first place?

Taking questions like these as a point of departure, in the next chapter I explore how the "materiality of interaction" might work as a theoretical notion for addressing interaction as an *activity* while simultaneously addressing the object(s) and elements at play during interaction.

3   The Materiality of Interaction
Understanding Interaction through a Material Lens

A concern for interaction

This book about *the materiality of interaction* is, obviously, concerned with how materials and the very notion of materiality are related to interaction design. But as we shift perspectives from representations to materials, we also need to look closely at this central notion of *interaction*. We need to have a clear idea of what it is, and consider whether a shift of perspectives also has implications for how this term should and can be used in the context of interaction design—and when designing *for interaction*. How do we define this term "interaction," and how do we see interaction as being at least partly a material matter? This chapter is devoted to these essential questions.

So what is *interaction*? This question seems short and simple, but is it really?[1] Of course, and in the context of human-computer interaction (HCI), we can say that "interaction is a turn-taking act between man and machine."[2] Such definitions of interaction are not uncommon across the history of HCI and interaction design. For instance, when HCI was to a large extent about command-based user interfaces, "dialogue based models of interaction"[3] were popular for thinking about this turn-taking in the form of a conversation between a user and a computer.

However, it is even more typical in our field that we focus on *the human, or a person*, and his or her interaction with a computer. For instance, in the now classical "Gulf of execution—gulf of evaluation" model developed by Don Norman (1986), we see a person going through these two gulfs when interacting with a computer. (And even though the computer is assumed to be part of these two processes, it is not explicitly addressed in the model.)

The field of HCI and interaction design has favored the human side of interaction in additional ways over many years. This is articulated in the *methods* developed in our field, including, for instance, *user-centered design* and participatory approaches to interaction design. But it is not only in the methods and approaches to interaction design that this focus on the human side of interaction is foregrounded. If we look at some of the well-established *theories* in our field, such as activity theory,[4] we notice a similar focus on the human as the core unit of analysis for understanding interaction. In activity theory, interaction analysis takes a point of departure in the human, the subject, who is using interactive technologies as mediating tools to accomplish a task or carry out an activity. Here the computer is a tool that works as a mediator between the subject and what he or she is trying to accomplish.

This "mediating structure" suggests something interesting, and opens up a whole new set of questions related to the nature of interaction. Maybe the tool is an entangled aspect of the interaction per se? Maybe the tool not only enables the interaction but actually defines it, scaffolds it, frames it into a certain form, and ultimately makes interaction possible at all? (Clearly, this role moves well beyond just "supporting" the interaction as if it could also happen without the tool.) As such, doesn't the tool-mediation perspective suggest that it is necessary to understand interaction in terms of an inseparable interplay between a person and some interactables[5] during sessions of interaction?

Even though "interaction" is actually a vital part of HCI, interaction is still a notion that we have missed or failed to address in detail. For sure, we have focused on the other parts of this acronym. We have focused on the "human" aspect not only through user-centered design methods (from formal and controlled experiments to more open-ended field studies, including ethnographic studies, etc.), but also by conceptually and theoretically addressing what the human aspects are from different academic viewpoints, including cognitive psychology, sociology, anthropology, and so on.[6] We have also carefully addressed the "computer." The whole field of computer science has been occupied with exploring how computing machinery can be further developed, and while there have been calls for even more focus on the "computer" in order to understand interaction (see, for instance, the book *Computers in the Human Interaction Loop*; Waibel and Stiefelhagen 2009), the entity of interaction is still not fully addressed.

So can we find any similar systematic attempts to address what interaction is—empirically and/or theoretically? In this chapter I will go through some of the most recent attempts, but first I will present some related recent work that has influenced today's integrated view on interaction instead of treating it as a separated "third notion" in the acronym HCI.

What do I mean when I say that we need to talk about an "integrated view on interaction"? In order to move forward, I would suggest, we need to move beyond this threefold separation of "humans," "technologies," and "interaction." In fact, thinking about interaction without thinking about either "persons" or "technologies" seems impossible. So can we find some conceptual ground for thinking differently? Can we find some stable ground for thinking about interaction as completely integrated with, and completely inseparable from, the people doing the interaction, and from the computational objects involved in interaction? In short, can we find a meaningful ground for thinking about and talking about the "materiality of interaction"?

This concept of "materiality" has been used in a number of different fields to talk about how different materials come together in a composition and to address how different parts are entangled in a larger whole. For instance, it is heavily used in the field of architecture to address the concept or applied use of various materials or substances in the medium of building. Here the various materials constitute the parts, and "the building" is the larger whole. In architecture, it is also typical to use the concept of materiality for thinking about architecture from the viewpoint of structural and aesthetic concerns.

In a similar way we can think about materiality in the context of interaction design. Here "materiality" can refer to the integration of physical and digital materials in the design of new interactive technologies and objects. Furthermore, we can think about the structural and aesthetic aspects of interactive solutions and can use the notion of materiality as a framing label for such concerns.

In the area of HCI and interaction design, the term "materiality" has its theoretical grounding and background in the work by Karen Barad and Daniel Miller. Karen Barad is an American feminist theorist with an interest in quantum physics who shares a concern similar to the one I am addressing in this chapter. She is concerned with an object of study which cannot be reduced to its parts, nor divided into a number of separate issues.[7]

Similarly, I want to avoid any division of HCI into the three separate aspects of *humans*, *computers*, and *interactions*.

When Barad (2007) introduced the notions of "materiality," "agential realism," and "entanglements," she managed to address a phenomenon without taking it apart, not even for analytical purposes.[8] Instead, she developed a vocabulary for addressing a highly integrated and inseparable unit of analysis, and she developed a language and a vocabulary for addressing, describing, and analyzing that unit of analysis as a whole. As I see it, we share a similar challenge here.

So, given this idea of the materiality of interaction, we might propose the following question as a starting point for moving forward: *What is "interaction" in the context of HCI if we are not simultaneously talking about human interaction with computational things?*

In order to start answering this question, I will use a simple example to illustrate how human actions and interactive technologies are inseparable in the production of interaction. Say that you are sitting in front of your TV and you want to switch to another channel. You reach for the remote and you push a button. You see that the TV is switching to a different channel, so you have instant feedback on your action. This is a simple example of interaction, involving a person interacting as well as an interactive technology—the TV set.

But what can this simple example tell us on a more general level? What is the core implication from this simple example for how we could think about interaction in an integrated manner in HCI? In our research,[9] we have suggested that what we need to develop is a *relational design language* for interaction design. If "materiality" is about how people and technologies are entangled in acts of interaction, then a relational vocabulary might tell us something about the interplay in this complex. We might not be able to map exactly how this interplay is configured—since it might not be simple connections or straightforward cause-and-effect connections—but this relational view will at least not overlook interaction while only paying attention to the people interacting or the particular interactive technologies used. In short, it will keep us focused on interaction per se (Wiberg 2010a), and on the flows of interactivity (Stolterman and Wiberg 2015).

If interaction design is not only about the design of interactive technologies, nor only about understanding user needs, but rather about designing the interaction that emerges between interactive technologies and the people using these technologies (in other words, how it includes the people[10] interacting and the configuration of interactive materials so as to support particular flows of interaction), then it becomes natural to think about this in terms of *the materiality of interaction*.

Of course, this is also true for other related fields of design. For instance, it is hard to talk about the field of service design only in terms of designing "services" without taking into account the people being served, the people providing the service, and the materials (products, etc.) that scaffold the service provided. Taking yet another example, the same logic can be applied to thinking about "experience design." In the same sense, it is hard to just design a particular experience, but with a particular experience in mind, one can try to design good preconditions for that experience to happen. For instance, having a clear target group in mind (maybe further expressed in the form of a persona), one can then design material resources in relation to that preferred experience and in relation to a particular user or target group.

In short, "the materiality of interaction" is a perspective that does not lean toward designing interactive artifacts first, and then think about people second. It is not a technology-driven design account. On the contrary, it is about making "interaction" the core focus, the point of departure and the core evaluation criteria, for thinking about what good interaction design is.

So, given that interaction is understood and approached as something inseparable from *what we interact with or through*, we can then think about the materiality of interaction as the unit of analysis and the core for interaction design projects.

Although this perspective works across the whole history of HCI / interaction design, I would like to clarify that while we are currently moving through a transition from representation-driven interaction design to material interaction design, this materiality of interaction is also going through a radical shift. In chapter 2 I talked a little bit about this in relation to how the notion of "direct manipulation" has shifted slightly from symbolic direct manipulation to even more direct (material) manipulation.

In HCI, this shift toward a new ground for interaction design has been labeled in many different ways in recent years, in addition to the "material turn." In the next section, I will address a couple of these notions, including "ubiquitous computing," the "third wave of HCI," and "tangible user interfaces."

Material interactions: On ubiquitous computing, tangible UIs, and third-wave HCI

The most profound technologies are those that disappear. They weave themselves into the fabric of everyday life until they are indistinguishable from it.
—Mark Weiser[11]

"Ubiquitous computing"—this term, coined by Mark Weiser, is indeed a strong concept[12] in HCI. It has proven itself strong not only because it serves as a good label for one of many current developments, but maybe foremost because it has worked as a marker for forward-thinking, future-oriented—and sometimes even speculative—interaction design projects.[13] Ubiquitous computing is a vision for where computing is going in the near future. As defined by Mark Weiser, ubiquitous computing is about computing "weaving itself into the fabrics of everyday life" to the point that the technology is indistinguishable from what it is entangled in. At this point, the technology "disappears" and becomes part of everything, and everywhere.

In relation to material interaction design, the idea behind ubiquitous computing is both informative and confusing. On the one hand, as I've just said, ubiquitous computing is about the "indistinguishable" integration of computing into our physical world. This fits in well with what I have described in this book as the tight integration of compositional interaction design thinking across physical and digital materials.

On the other hand, the notion of ubiquitous computing seems to lead us away from material interaction design. When computing becomes so integrated that it "disappears," it also loses any physical form; formulated slightly differently, in order to "disappear," it needs to lose form as a critical existential criterion. The ultimate example is "cloud computing," i.e., computing in the cloud, free from any physical form and accessible from anywhere, rather than thinking about how computing manifests itself more and more in physical material forms.

While Hiroshi Ishii's vision of tangible user interfaces (TUIs) seems to show the way forward in terms of how computing can get more integrated in everyday objects, it represents a quite different perspective in relation to the "disappearing" character of computing that Weiser describes. Ishii probably shares Weiser's vision of integrating computing power in everyday objects, but the difference is that Ishii wants to explore alternative ways for computing to be manifested in material form—in contrast to making computing completely disappear.

Still, both of these perspectives are important for understanding the materiality of interaction. Interactive technologies are increasingly ubiquitous. We carry computational power anywhere we go, we have access to the Internet from just about any location, and we have computational power embedded in everyday objects to the extent that we have stopped noticing them. Who thinks about the computers embedded in our ID cards (those with RFID tags or smart chips)? Beyond that, who thinks about the use of these ID cards mainly from the viewpoint of human-computer interaction? Probably very few people do, and those few are probably people who are doing interaction design themselves.

A third perspective that has guided recent developments in our field is the trend toward "third-wave HCI."[14] In short, the third wave in our field focuses on how the use of computers spreads from the workplace to our homes and everyday lives and culture. As such, the "third wave of HCI" is about ubiquitous computing, but from the perspective of the user rather than from a technological perspective. Third-wave HCI looks at how computing, not only as a technology but as an activity, is becoming completely blended not only with everyday objects but also with just about any everyday context, social context, and cultural practice.

In relation to "the materiality of interaction," this notion of third-wave HCI tells us that the context for interaction is not only the thing that sets the stage for the interaction, but also the thing that to a large extent affects how the interaction unfolds. In short, in order to understand the materiality of interaction at the current moment, one needs to take into account this third wave of HCI.

Before going into the current frameworks and models developed for this notion of materiality in the context of HCI / interaction design, I will touch upon what the literature can tell us about interactivity and materiality in the area of interaction design research.

On the notions of "interactivity" and "materiality" in interaction design research

At the beginning of this chapter I said that there are two ideas about *the materiality of interaction* that need to be elaborated. First, we have the notion of interaction or "interactivity," and second that of "materiality" and, in particular, how it has been used in the context of HCI and interaction design. Here I would like to elaborate a little on these two strands.

Interaction or "interactivity" is not easily defined. In the world of analog objects, we can of course interact with objects with no computational power. For instance, if I heat up my grill, I can interact with it while barbequing. I can observe temperature changes, I can take different actions to adjust the temperature, I can move the food on the grill around to make use of hotter and cooler parts of the grill, and so on.

This example leads us to an obvious question: Is there a fundamental difference between interaction with physical objects and interaction with digital objects? And what about physical objects that also have some computational capabilities? And by the way, how much computational power does an everyday object need to have in order to be understood as a computational object? While it could be interesting to think about how digital objects might be ontologically different from analog objects, any such distinction is increasingly blurred as we move toward a material-centered interaction design practice where this distinction between atoms, bits, and even cells is increasingly blurred or eliminated. So, given that the analog and the digital, or the physical and the virtual, might soon be indistinguishable from each other, it might actually make more sense to talk about what interaction or interactivity is, rather than having categorical concerns about what kind of substrate can be said to be part of computing or something that is "interactive." Just think about the Moleskine Evernote example introduced earlier in this book, where a number of computational and traditional objects make the interaction design complete. While the Moleskine Evernote app is to a large extent digital, the notebook and the pen are physical. It is the composition of a notebook, a pen, a mobile app, and the Moleskine Evernote cloud service that makes the interaction design complete, and it is that interplay of materials that taken together configures the scene for the interaction that it can support.

Returning to the literature on HCI and interaction design, we should notice that as a field we have in fact conceptualized "interactivity" in a few different although related ways. One of the first attempts to conceptualize interaction, that of Löwgren and Stolterman (2004), provides the important cornerstone that we can use to think about interactive artifacts not only in terms of a user interface and input/output modalities but also of an interactive system's "dynamic gestalt." Löwgren and Stolterman suggest that through interaction an artifact reveals its dynamic gestalt. For instance, a classic arcade-style computer game reveals different levels, monsters, obstacles, and challenges along the way as the user continues to interact with the game. The same goes for a simple ATM machine, where the dynamic gestalt is presented through a dialogue between the user and the machine. Here the flow of interaction guides the dialogue to, say, a cash withdrawal or some other service offered by the machine.

In short, this notion of dynamic gestalt tells us that interactivity is not merely about the turn-taking between a person and a computer. In addition, the "interactiveness" of an interactive system, a digital service, or a so-called "smart" object is a quality that can be discussed in terms of the "dynamic gestalt" of an interactive system, and how this gestalt allows for the presentation of the interactive system, object, or service over time.

With this conceptualization, Löwgren and Stolterman navigate away from seemingly simple yet more problematic definitions of "interactivity" as something that needs to have a (in most cases visual) user interface. In fact, Janlert and Stolterman (2015) have described "faceless interaction"—interaction with interactive artifacts that do not have a visual interface (for instance the Amazon Echo system, which only has a voice-based interface).

Similar to the notion of "dynamic gestalt," Vallgårda et al. (2015) have recently introduced the notion of "temporal form." This concept is also about interactivity from a temporal perspective, but instead of focusing on the flows of interaction, temporal form is about the visual and material presentation of the interactive artifact over time. According to Vallgårda et al.:

Interaction design is distinguished from most other design disciplines through its temporal form. Temporal form is the computational structure that enables and demands a temporal expression in the resulting design. When programming computers we create a temporal form that then comes to expression through an output of actuators and other materials. Indeed, it is these material manifestations of temporal forms that enable our interactions with computational things, as digital computations in themselves are inaccessible.

Again, and similar to Löwgren and Stolterman, Vallgårda et al. seek to develop a vocabulary for talking about interactivity in relation to interactive qualities of computational objects. Both focus on these qualities of the object to the point that it seems impossible to understand interaction without at the same time understanding interactivity; this return in focus leads us back to reviewing the current discussion on materiality in our field.

"Materiality" is a term that has quite rapidly established itself in the area of interaction design. As described earlier in this chapter, "materiality" can refer to the integration of physical and digital materials in the design of new interactive technologies and objects. We can think about the structural and aesthetic aspects of interactive solutions and can use this notion of materiality as a framing label for such concerns.

Over 200 papers have been published in the last five years that use this notion within the field of HCI and interaction design. Some of the most referenced include works by Jung and Stolterman (2011) on digital form and materiality, Giarcardii and Karana (2015) on material experiences, Sundström et al. (2012) on immaterial materials, Vallgårda and Redström (2007) on composite materials, Pierce and Paulos (2013) on "electric materialities," Dourish and Mazmanian (2013) on the materiality of media, Wiberg and Robles (2010) on the material turn, and Dourish (2015) on the materiality of information infrastructures.

Indeed, our community has been quite successful over the last few years in developing concepts that address various aspects of the movement toward a material understanding of interaction and interactive systems. But the development of concepts can best produce a vocabulary for talking about different aspects of this phenomenon.

Beyond that, there has also been some work devoted to increasing our understanding of the connections and interdependencies among these notions. We have seen some great attempts made to develop theoretical frameworks that specifically focus on this notion of "materiality" or "digital materials" in the area of interaction design. The next section presents two of these frameworks and discusses how they add to our understanding of "the materiality of interaction."

The materiality of interaction—Current research, frameworks, and examples

The third wave of HCI—including the movement toward ubiquitous computing and the development of more tangible ways of interacting with and manifesting computing in material form—does not only call for new frameworks for doing interaction design and new design ideas within material-centered interaction design thinking. It also calls for new ways of conceptualizing and theorizing this movement, and for an understanding of how the basic notions of interaction, materials, and forms are interrelated.

Motivated by this trend, Jung and Stolterman (2011) have pinpointed that form and materiality need to be posed as a foreground concern for interaction design. In their work they have focused on the form and materiality of interactivity, and in doing so they have proposed a framework that specifically addresses the interrelations between an artifact, its users, and its designer. The framework also outlines how an artifact can be thought of both from a design perspective, in terms of its form (including how it relates to materials used, its shape, and the process of making it), and from a use perspective. In the case of an artifact's use, the notion of materiality addresses not only how it is used but also how that use relates to processes of sense making (meaning), use patterns (using), and the artifact's relation to other artifacts (ecology). As Jung and Stolterman (2011, 648) describe these interrelations:

Materiality covers a broad range of values and effects that are elicited in the course of making and using material artifacts, broadly indicating the relationship among an artifact, people, and/or other artifacts.

These interrelations are further expressed in their model of how form and materiality are related to artifacts; see figure 3.1.

Further on, Jung and Stolterman (2011, 648) suggest:

While form is about design considerations with its three conceptual dimensions of material, shape and making, materiality is more about the relationship between people (user) and the material artifact in terms of how it is used out in the world. In close relation to how an artifact is formed, materiality can provide useful perspectives to investigate aesthetic and experiential qualities of digital artifacts.

When they say that "materiality is more about the relationship between people (user) and the material artifact in terms of how it is used out in the world," their assertion fits in perfectly with my notion of "the materiality of interaction." The core argument in this book is that it is the materiality of interaction (not the materiality of the artifact) that is the most interesting aspect, and yet the most overlooked aspect, of interaction design. Any interaction designer needs to stay focused on the interaction that is going to be designed (and then how particular artifacts can be made to realize that particular kind of interaction). So, if materiality is, in the words of Jung and Stolterman (2011), a "relationship between," then their model

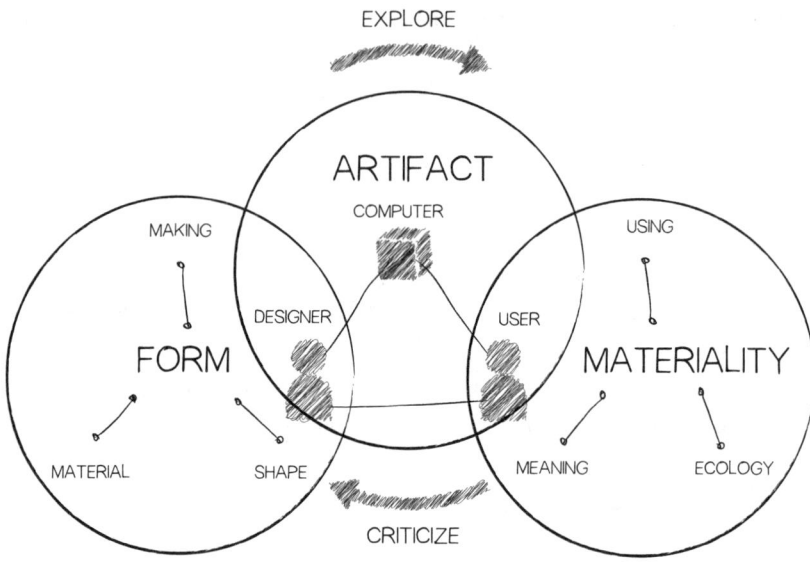

Figure 3.1
Model showing how the notions of form and materiality are related to artifacts, the
use of these artifacts, and designers as makers of these artifacts. From Jung and Stolterman (2011).

is a good argument for keeping this focus on the "in-between"—i.e., on
"interaction"—when thinking about materiality in the context of interaction design. In addition, their model addresses the "in-between" aspects
without excluding the material artifact. Again, this fits perfectly with my
argument that a focus on "the materiality of interaction" can be combined
with a method for doing material-centered interaction design, which I will
elaborate on in chapter 4.

While Jung and Stolterman's framework is very clear about how materiality relates to artifacts and to the notion of "use," there are additional
dimensions to take into consideration as we move through the third wave
of HCI. For instance, as interaction designers, we do not only focus on how
a digital artifact should work in order to support a particular type of use; we
are equally concerned with how it is experienced (thus the notion of experience design). Some frameworks have recently been developed that help
to navigate how we can think about the relation between users (people)
and material artifacts (materials) from a more experiential viewpoint. For
instance, Giaccardi and Karana (2015) have developed "the materials

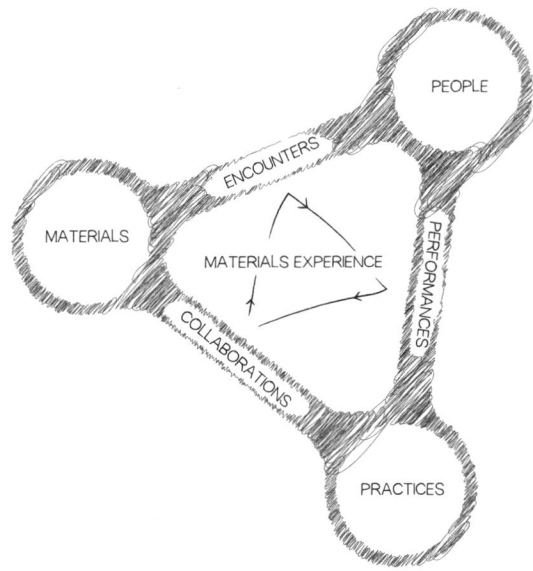

Figure 3.2
An illustration of the materials experience framework developed by Giaccardi and Karana (2015).

experience framework," which acknowledges the experience people have when interacting with and through materials. As they explain their definition of the notion of "materials experience":

"Materials experience" is a phrase that acknowledges the experience people have with and through materials. Originally, the expression was coined to acknowledge the active role of materials in shaping our internal dialogues with artifacts. However, we argue, a comprehensive definition of materials experience should acknowledge also the active role of materials in shaping our ways of doing. In other words, it should attend to the aesthetic aspects of experience as much as to its performative character. The proposed framework offers, therefore, a bridge between the theoretical roots of experience design as already established in the HCI field, and the growing practice orientation of HCI accounts of materiality. (Giaccardi and Karana 2015, 2449)

So in short, they present a way of uniting our community's focus on materiality with our already established accounts for doing experience design, and they present it in the form of a model (see figure 3.2).

According to Giaccardi and Karana (2015, 2454), their proposed framework provides designers first with a vocabulary for articulating and configuring the situational, experiential whole in which materials and practices are implicated. Second, it helps illuminate and frame future research at the convergence of materiality scholarship and emerging practice-oriented agendas in HCI, by connecting important understandings produced within these strands of research to the theoretical roots of experience design as already established in the HCI field.

Thus, we not only interact with these artifacts (in terms of "use" and how we make sense of new interactive artifacts, that is, their meaning), but we also experience these interactive materials ("material experiences") via a cultural and social lens, here referred to under the label of a certain "practice."

To return to our initial question: *So what is interaction?* Well, if we take into account the dimensions addressed in these frameworks, I would repeat that interaction cannot be understood only as a turn-taking act between a user and a computer (as already stated in chapter 1). Beyond that, *the materiality of interaction* deals with an interplay in a context that Giaccardi and Karana (2015) refer to as a "practice." Furthermore, the materiality of interaction is inseparably related to a *person* using an interactive artifact. As such, this notion of materiality provides us with an applied focus on interaction and interactivity, in the same way that materiality is used within the field of architecture for speaking about "the applied use of various materials."

Interaction is, additionally, about how people interacting with digital materials make sense of them and find them meaningful, how they use them (use patterns, particular ways of using them, routines, workarounds, in cases of breakdowns, etc.), and the ecology that the digital materials are part of. Furthermore, not only is the materiality of interaction experienced in a particular context, in a particular practice (including an understanding of practice from a material, cultural, social, and historical perspective), but material experiences—i.e., how the materiality of interaction is perceived—are also formed through the different ways the materials of interactive artifacts are presented to us, how we encounter these materials, and the experiences we gain from these encounters.

In short, "interaction" is indeed a *relational concept* in itself, and not just a concept for addressing relations.

In the next chapter I will discuss the philosophy behind material-centered interaction design, as well as a methodology for doing it. I will also present ways to address, investigate, and work with a material focus in interaction design projects, illustrating how this approach offers a complementary design perspective that opens up a design space beyond the categorical division of skeuomorphic versus non-skeuomorphic interaction design. In moving beyond any such dichotomy, I will address how interaction stand in close relation to the tradition of craft, and I will explain how design thinking, influenced by this tradition and practice, can enable us as interaction designers to work even more seamlessly across digital, analog, smart, physical, and even immaterial materials—in fact, to work in interaction design across any kind of substrate.

4   Material-Centered Interaction Design
Toward a New Method and Foundation for Interaction Design

The growing interest in the materiality of interaction in the field of human-computer interaction[1] indicates that there is a value in acknowledging *the material aspects and dimensions of interaction design*. However, if we rely only on a representation-driven approach to interaction design, the notion of materiality only works, at best, as a "metaphorical maneuver,"[2] and we would still advance an interaction design paradigm oriented toward the immaterial aspects of interaction design (for instance, the use of symbols and metaphors in interaction design).

What would an alternative perspective and approach be? Can we not only shift perspectives here, but also imagine different approaches to and methods of interaction design that would truly accept the digital as a design material, would focus on interaction as the form being designed, and would not introduce any categorical distinctions between different types of materials? In this book I have suggested that we should make no metaphysical or ontological distinction between physical and digital materials, between atoms, bits, and cells, or between "visible" or "invisible" materials, and that we should even avoid distinctions between what might be considered as "material" or "immaterial" in the first place (like radio waves). In the same way that wood and iron are typical examples of physical materials, I consider code, algorithms, sensors, and processors as digital materials. Still, from the viewpoint of interaction design, it is the composition and activation of these different materials so as to give the interaction a particular form that is essential—not each material's ontological or metaphysical status. So, instead of focusing on what a particular interactive system represents, the material-centered approach to interaction design as proposed in this book focuses on how *interaction* is *presented* and *materially manifested* in the world (in all imaginable forms—from completely embedded and "invisible" interactive systems to the different forms of gadgets we surround ourselves with in our everyday lives).

In this chapter I develop this idea into an approach to interaction design that I label *material-centered interaction design*.

As discussed in chapter 1, it sounds almost like a paradox that something so seemingly immaterial as interaction can be designed through a material-centered approach. Still, with the growing design trend toward ubiquitous computing, the Internet of Things, and embedded systems, interaction design is increasingly manifested not only as purely digital services or in "the cloud" but also in physical form. If we are now treating physical and digital materials as fundamentally similar substrates (on a metaphysical level), a material-centered approach to interaction design considers both (1) how different physical and digital materials are brought into composition in the design of interactive systems, and (2) how code, computing, algorithms, scripts, variables, functions, protocols, databases, networking, and information sources are interaction design materials that need to be brought into composition with other materials so that interaction design can be manifested in objects, devices, vehicles, and other instantiations of interaction design in our world.

To give one simple example, a smartphone today is not only a small computer in a glossy shell; it is also actually "smart" in many different ways. First of all, it is smart in the sense that it is not limited to being a complete and "closed" product. Instead, app stores and similar marketplaces enable its owner to configure, personalize, and update the smartphone time and again over its lifespan. From this viewpoint, the smartphone is "open" to change and adjustable in relation to shifting user needs.

A smartphone is also smart in the sense that it can distinguish human fingers from other materials touching the device.[3] Touch enables the user of a smartphone to "touch the information on the device" and to operate its user interface. Through this smartness, the device's response to touch mirrors the distance between what the user wants to do with the device (intention) and how that is realized (action). In short, touch shrinks "the gulf of execution/gulf of evaluation cycle,"[4] enabling users to quickly move from an intention to "touch" a piece of information to actually doing so.

In addition, most smartphones today are location-aware. For instance, if you play Pokémon GO or if you use Google Maps on a smartphone, the app will know your location and accurately position you/your device on that particular position on the map. Today, this location awareness is used for different map applications, exercise apps (for instance the Runkeeper app), games, and speedometers. A "smart" device in the form of a physical device that is aware of its current location enables interaction design aimed at aligning physical geography with interactive systems design; even something as "immaterial" as a position, a speed, or a distance can work as a design material in interaction design.

Finally, to return to the beginning of this example, a smartphone is not only a mobile interactive system; it is also a physical device, a gadget, a thing, or "an object." For this reason, it matters how this interactive device has been physically manifested in terms of its form factor, weight, size, color, material choices, and so on. When we see someone using a smartphone, it is not the apps that we see first. Rather, we see its physical manifestation in the public space—the device. Therefore, the interaction design of smartphones is also to a large extent about physical materials, and interaction design is from that viewpoint also a matter of classical industrial design.

These are some examples of aspects of a smartphone that its users experience in their everyday use. All of these aspects are manifested through the integration of different materials in the design. The materiality of smartphone interaction can therefore be said to be defined by these aspects, as well as the additional ways the interactive device has been realized in terms of its design materials.

In this simple example of the smartphone, we should notice how different materials come into play to realize the user's experience of it. Modern smartphones rely on a range of technical systems to work in concert, including touch (and nowadays even pressure sensors to enable 3D touch) and algorithms to distinguish a quick touch from a swipe across the screen. The model of the app store, which allows the user to update and change the functionality of the smartphone, is also enabled by a multitude of different materials coming together. App stores demand a whole range of subsystems to work, including phones that are upgradable and an OS that can run different applications on the phone. Furthermore, app stores require working network connectivity, platforms, and a functional ecosystem (including technically sound solutions for secure payments, app updates, bug and malware checking, security, etc.).

Finally, we can think about the materiality that enables a smartphone to appear to its user to be location-aware. Location awareness is typically implemented on smartphones by relying on GPS technology. GPS is in itself a complex system of integrated solutions including a GPS chip on the phone and, more fundamentally, satellites, triangulation algorithms, and a worldwide standard to make it work. The materiality of experiencing a "location-aware" smartphone is thus a matter of a multitude of technical and material solutions working in concert. Consequently, interaction design becomes a matter of material considerations despite the fact that the interaction per se might be thought of as highly immaterial.

The smartphone is just one example of the tight connection between how we experience an interactive system and its enabling technologies; this experience is not a one-to-one relation between an experience and a particular technology, but rather is the result of the deliberate configuration of many materials working in concert in the enabling of and experience of an interactive system. This example illustrates that (1) many different materials need to work in concert with one another, (2) different materials are central in the design of interactive systems, and (3) interaction is always dependent on the materials enabling the interaction. In short, it demonstrates the importance of a material-centered approach to interaction design.

I would like to argue that *material-centered interaction design* is an alternative to the metaphorical way of thinking about the materiality of interaction. Beyond thinking about computing and information as abstract design materials, it is about focusing on materials when designing interaction.

As the smartphone example illustrates, this material-centered approach to interaction design does not neglect a user focus when designing interaction. Even though it might sound counterintuitive, focusing on materials in the design of interactive user experiences is a *relational approach* to interaction design that has one keen eye on user needs and another on materials that can enable a preferred interaction design. Interaction is then designed through material configurations. Without a detailed understanding of the materials, the interaction designer might not arrive at the intended interaction design.

In short, a material-centered approach to interaction design is all about how to manifest interaction in material form, and how to best utilize different materials in the design of particular and preferred modes of interaction.

Addressing materials: Moving beyond non-skeuomorphic interaction design

Other approaches to interaction design encompass this notion of "materials" but in a metaphorical way. In particular I am thinking about the current discussion on "skeuomorphic interaction design" versus "non-skeuomorphic interaction design," and in this section I will distinguish the *material-centered approach to interaction design* from both of these other strands of thought.

The main reason for making this distinction is that there is today a trend toward non-skeuomorphic interaction design in the area of interaction design paradigms and de facto standards on the design of GUIs (graphical user interfaces).[5] The overall message of this trend is that we have reached a moment in the design and development of digital services and products where we no longer need to rely on either metaphors or a visual linkage to the physical world in order to make good and useful interactive systems. In short, the detour via the physical world is not needed in order to understand the digital world—and thus the world of digital design.

Now, if we take a look at these three approaches to interaction design, we can say, at least on a general level, that all three of them relate to "materials" but in three distinctly different ways:

Skeuomorphic interaction design: UIs influenced by physical materials. Here the idea is that the digital does not have any inherent material qualities, and needs to borrow these qualities from materials and objects in the physical world. Some people have recently expressed this by stating that skeuomorphic interaction design relies on a physical heritage.[6]

Non-skeuomorphic interaction design: no material influence. This approach is quite the opposite of skeuomorphic interaction design. It holds that digital materials have their own properties, and so digital design does not need to resemble something physical. In short, instead of borrowing its properties and form from the physical world, digital design is developed without any such linkage to the physical world.

Material-centered interaction design: from "influenced" to what the UI is actually made of. Finally, and in contrast to both the skeuomorphic approach, which builds on a physical heritage in digital design, and non-skeuomorphic design, which avoids any linkage to the physical world, the material-centered approach as proposed in this book does not uphold any distinction between the physical and the digital in the first place. Here the idea is that we need to consider interaction design as composed of both digital and physical materials. Accordingly, such interaction design cannot be fully non-skeuomorphic in terms of its digital form (due to the properties of the physical materials used in the design), nor can it be merely skeuomorphic (in terms of resembling something physical). Beyond any such metaphorical connection, interactive systems built on both digital and physical grounds are referred to as "material interaction design" in this book. A material-centered approach to interaction design moves beyond interaction design that seeks to "resemble something" toward interaction design that actually "is something."

When I say that interaction design "is something"—as in being a thing in our everyday world—I mean that it (1) manifests itself in material form, and (2) utilizes the combination of physical and digital materials for that manifestation. Let me introduce two examples to further illustrate this point.

The first example is the activity band Misfit Ray, a Bluetooth-equipped step counter in the form of a bracelet. Besides the Bluetooth chip, the device contains a tiny computer, an advanced accelerometer, and a couple of batteries. The main function of this device is to work as a step counter that syncs with your mobile phone, but it can also work as a simple remote for any number of different smartphone apps. Its user can interact with the device either by tapping on it or by accessing it via a smartphone app. For instance, by gently tapping the bracelet, the user can switch to the next song in a playlist. However, the form factor of Misfit Ray—and its materiality—does not resemble a traditional computer or even a smart watch, as the device does not have a screen or any buttons. From a form factor perspective, it is only a bracelet, as illustrated in figure 4.1.

By deliberately focusing attention on *the bracelet* as the overarching form factor of this interactive device, the device "hides" its different modes of interactivity and functionality.[7] Still, the device can do a number of different things (for instance, work as an alarm clock and vibrate to wake you up in the morning, vibrate at a chosen interval to remind you to move now and then during a work day, monitor steps taken and hours of sleep). Yet another thing that makes the computational aspects of this device

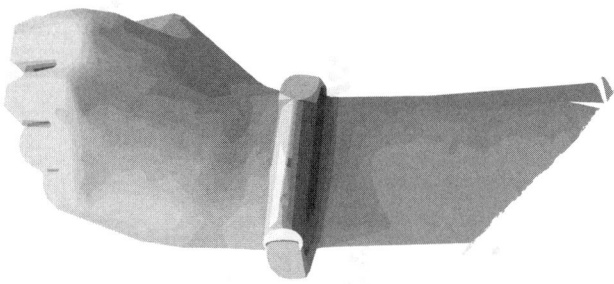

Figure 4.1
The Misfit Ray bracelet.

"disappear" (besides the lack of a glowing screen and buttons) is that it only needs new batteries every sixth month. This makes it unlike almost any mobile device nowadays, for which daily charging is a typical, although maybe not preferred, part of our everyday interaction.

In short, the interaction design of the Misfit Ray has been deliberately devised to make the computer disappear, while simultaneously making sure that this computational object is manifested in physical form[8]—it's an example of *material interaction design*.

Another example is the Amazon Echo device. The Amazon Echo is a WiFi-connected and voice-controlled smart object with the form factor of a cylinder. This cylinder form and the size of the device are very similar to the form and size of many small Bluetooth speakers (see figure 4.2 for an illustration of this form).

Similar to the Misfit Ray, the Amazon Echo does not have a screen or any buttons. It even lacks an on/off button because part of the interaction design is that it should never be switched off. Being "always on" is an important aspect of the device's materiality because it is always ready to interact with its users via voice-based questions and commands. The Amazon Echo can answer its user's questions (for instance, questions about current news, being asked to Google things or to play music), and since it also knows its location, it can tell you the local time and local weather, for example. The Amazon Echo has this functionality because the cylinder hides a computer inside; it is also equipped with Internet connectivity and cloud-based digital services.

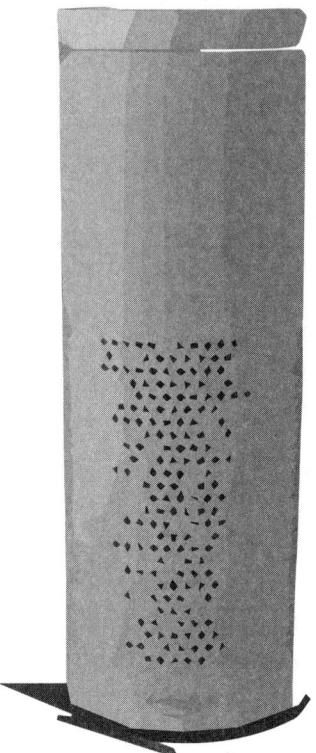

Figure 4.2
The Amazon Echo.

Some clear similarities stand out between the Misfit Ray and the Amazon Echo. Both rely on advanced algorithms to work; for instance, the Misfit Ray relies on advanced algorithms in order to recognize walking as an activity different from typing, and the Amazon Echo can distinguish voice commands and oral questions as distinct commands in relation to other conversations held in the same room. In addition, both devices illustrate alternative ways of interacting with computers beyond the traditional keyboard, mouse, and screen; and both rely on the tight integration of physical form, computing, network connectivity, and advanced algorithms in their design. In other words, both of these examples illustrate compositional and material-centered interaction design across physical and digital materials.

These two devices also illustrate interaction design that is deliberately designed to blend in, either in the form of an accessory you wear or as part of your home. Again, the material instantiation of the preferred interaction—that is, the materiality of interaction—is carefully considered in these two examples of material-centered interaction design.

Approaching materials and material properties—Craft and design

But how do you go about creating such a design? Is there a particular approach that you can rely on in doing it?

Instead of inventing yet another novel approach for careful design, it might at this point be good to stop and reflect on the existing accounts of careful design in close relation to the materials at hand. That is, we can seek a careful approach to interaction design that can also work as an approach to doing material-centered interaction design. Here, the growing interest in craft-based approaches to interaction design surfaces as an interesting candidate.[9] Craft is a material-centered way to give form, and is also a careful approach in terms of the outcome of the design process, how the final product is designed, and how the material is treated during the process. Good craftsmanship is to a large extent a careful process in which different material properties are carefully considered and utilized in the process of design.

Even though craft has its roots in other material-centered design traditions (not least wood, iron, and glass), it still bears an interest for a material-centered approach to interaction design. This is because craft is based in a sensitivity to materials: a sensitivity that includes attention to details and a developed sensitivity to material properties.

I do not see craft and design as synonyms, nor as mutually exclusive. Instead, I view them as complements to each other: I see design as an overall approach and process, whereas I view craft as a specific attitude toward and way of working with materials in the design process. From that viewpoint, a focus on craft underscores the fact that a design needs a good understanding of the materials at hand.[10]

Different materials—Digital, analog, and smart materials

Are we not facing an obvious paradox here? Why should we talk about different materials in a book that argues against making a distinction between physical and digital materials? And why distinguish between different materials in the first place? (Might it be the ultimate consequence of acknowledging particular material properties?)

What is important to remember is that a material-centered approach to interaction design does push for looking at interaction design through a material lens. In doing so it makes no ontological or metaphysical distinction between digital and physical materials. However, saying that the digital also counts as a material does not rule out the possibility of then distinguishing between different materials. That is still a material understanding (in comparison to not understanding the digital as a material). Instead, everything matters, but difference matters!

As examples in this chapter have already illustrated, physical materials can be activated to work as input/output to computing, and accordingly, interaction design stretches across different materials. Even "immaterial materials" (such as location, speed, or radio waves) can cooperate with digital and physical materials in the design of interactive systems. However, in order to design good material compositions, we must know the specific properties of different materials. If we know that we can build a rich repertoire of interaction design materials, we can develop skills to work with different materials in interaction design projects. Furthermore, the more we know about each material, the better we can become at working compositionally across different substrates. This is important in order to do good material-centered interaction design.

Different materials—Different material properties

"Material-centered interaction design" is not only about interaction design manifested in physical form. Doing "material-centered interaction design" also implies having a certain attitude toward the design work and the design process. This attitude means seeing and utilizing different material properties in design projects, and having the ability to see how different material properties might be useful in the design of interactive systems. Whether being "material-centered" refers to a desire to manifest interaction design in physical materials, or to a particular design attitude, for instance to approach it as craft, it is important to have a good understanding of different design materials. Here such materials can be categorized as digital materials, traditional/analog materials, and new/smart materials.

Digital materials (computational materials)

Let us first consider digital materials.[11] In this book, "digital materials" refers to the materials that computer *hardware* is made of (including the full range of abstractions from the raw materials of chisel, plastics, and copper to the pieces of any computer: integrated circuits, memory, cooling fans, circuit boards, wires, etc.), *software* (including the "stuff" software is made of, including bits, code, and scripts), and *infrastructures* (including again hardware like antennas, network cables, routers, etc., as well as enabling software for infrastructures, and not least the protocols that enable data to be sent across these infrastructures).

At the moment, there is also an increasing interest in the development of new and highly advanced sensors (including camera sensors, range sensors, and fingerprint sensors) capable of "sensing" aspects of the physical world (including light conditions, location, temperature, sounds/noises, and so on). These sensors also count as digital materials, and from the viewpoint of this book, the increasing interest in using such sensors in interaction design further demonstrates that any separation between the physical and digital world is not just conceptually difficult to uphold, but also on a practical level increasingly meaningless. Interaction design is increasingly about *weaving together* the world of bits with the world of atoms.[12] As formulated by Hiroshi Ishii: "At another seashore between the land of atoms and the sea of bits, we are now facing the challenge of reconciling our dual citizenships in the physical and digital worlds."[13] This "weaving together" can happen in many different ways, and typically a digital artifact is "woven together" on many different levels. Beyond the combination of hardware and software, and beyond the need for a physical (in most cases, plastic) container for electronics, the "weaving together" is also very much about the code, the scripts, the sensors, and the algorithms that, taken together, and through their particular configurations and integrations, enable artifacts to be interactive, and thus enable interaction. We should also not forget about the "immaterial materials" that more and more frequently serve as a backbone for many applications (including positioning data, tracking, radio waves, and so on).

Traditional/analog (traditional) materials

Almost in direct contrast to these digital and "immaterial" materials we have the traditional analog materials. As I pointed out in chapter 1, we have a history in the field of human-computer interaction (HCI) of designing and building computational machinery by combining digital materials (for instance, circuit boards and other pieces of hardware) with traditional physical materials (for instance, the use of wood as part of the first computer mouse). However, beyond HCI's history of working with traditional analog materials, we should not forget the much longer history of the design field's close work with particular analog material, and how one such material-centered focus has led to very unique strands of traditional design.

For instance, a number of well-known designers have made it their signature to design with a focus on a particular material. Le Corbusier had concrete; Mies worked with steel, glass, and marble; and Alvar Aalto worked with the skinning, steaming, and bending of wood. Similarly, the industrial designer Jonathan Ive, currently chief design officer of Apple Inc., has made it his signature to work closely with digital materials (including novel UI design solutions), as well as with the digital in combination with physical materials such as plastics, aluminum, glass, and most recently ceramics. Ive's description of his work on the physical "Jet Black" aluminum cover for the iPhone 7 serves as a useful example of his attention to material details (see figure 4.3).

The design of the iPhone 7 case illustrates the value of working with traditional materials as part of an interaction design project. It is also a good example of this book's focus on working closely with the details and the properties of a traditional physical material as part of the design of an interactive product. According to fastcodesign.com, Apple worked on the aluminum case for the iPhone7 in a very particular way, using "rotational 3D polishing, a protective oxide layer, ultrafine iron particle bath" to create a "pristine, mirror-*like surface.*"

From the viewpoint of industrial design, this describes the manufacturing process of the case for the iPhone, but it is also a particular design process aimed at achieving a particular look and feel for this interactive product. While one could argue that appearance and feel are mostly issues for industrial design, I would say that the physical dimensions of an interactive artifact also need to be well aligned with the digital design of that object. As discussed earlier in this chapter, crafting the digital artifact through its various materials constitutes the manifestation of the product. And, given that the iPhone 7 case is well-crafted—including the crafting of the relations between the materials—it is also likely that the artifact is interpreted and understood as a functional and integrated whole.

Figure 4.3
The "Jet Black" iPhone 7 case.

This example illustrates how the "look and feel" of an interactive artifact is no longer only about graphical design, nor is this merely a term used when discussing how easy it is to interact with a digital artifact. Beyond that, this notion of "look and feel" is increasingly extended to account for the whole artifact, and attention to details—all the way across a digital artifact (including its physical manifestation)—is central in order for the artifact to be perceived, valued, and used in the intended way.

Additionally, this example illustrates how one can pay close attention to (physical) materials not only through theoretical or observational study of the material, but by exploring and studying how the properties of the material enable it to take new forms, expressions, and behaviors through the process of being changed, refined, and polished. As I have argued, it is through that close and engaged relation to the materials that we can design and craft new (computational) things.

Other traditional/analog materials that could be explored in similar ways in the context of material-centered interaction design include wood, concrete, glass, and ceramics, just to name a few. In fact, there are some projects under way now that explore the use of concrete in interaction design, while others are exploring not only concrete and electronics but also their combination with wood in the design of everyday objects such as interactive alarm clocks.

New/smart (dynamic) materials

Interaction design is not just limited to digital design, nor is it limited to digital design in combination with physical design. Over the last few years, we have seen many attempts to design new smart and dynamic materials, including memory shape alloys, conductive fabrics, and thermochromic ink.[14] Indeed, the area of "material science" and the industry that is producing these new materials have opened up very interesting new opportunities for interaction design. As formulated by Wiberg (2016, 1201):

From water drops and soap bubbles, to conductive fabrics, thermochromic inks, bioplastics, and artificial grass, there is no shortage of proposed expansions to the material catalogue for interaction design. This explosion of interest in digital-physical substrates is a testament to the extraordinary success of ubiquitous computing; computing has moved out of "the box" and into the world. As the interface relocates from the glowing screen to the physical substrate new visions for the future of HCI are being proposed. For example, Ishii's "radical atoms" imagines a future in which all digital interactions are conducted through material substrates and thus human-computer interaction gives way to human-material interaction.

These new materials are interesting for a number of reasons. First, while traditional materials can respond to external commands (for instance, a servo motor can "bend" a piece of wood), these new materials can shift shape, color, location, and so on without any external command or force. At the current moment such opportunities have sparked interaction design innovation with an orientation toward "shape-shifting user interfaces."[15] Secondly, these materials open up new opportunities for expressing information as well as for changes in interaction in a material form (beyond the presentation of data, information, or the system status as glowing pixels on a screen).

But finally, and not unimportant in this context, even though new smart and dynamic materials are able to do or express new things, we need to move beyond any kind of one-to-one relational idea about what a particular material can do as part of an interaction design (for instance, the fact that wood can be bent illustrates this relation between a material and something its material properties allows or enables it to do). Furthermore, we need to move beyond the idea that a particular interaction design project can include only one material (physical or smart) as its manifestation.

Instead, interaction design happens today through the combination of different substrates and through its integration across many different substrates. For instance, when a cloud service is combined with a mobile fitness app, "running" is both the activity undertaken and at the same time the main interaction modality: the pace, distance, and route of the run not only constitute the experience of the runner but are also input to the system, and become the source of data for visualizations of the run and reflections afterward.

However, in order to move interaction design into our everyday lives and everyday environments, we need some guiding ideas not only about what characterizes different materials, but also about the design process for doing material-centered interaction design. This is the topic of the next section, and also a topic I will revisit later in this book.

On doing material-centered interaction design—The "interaction-first" principle

If computing is no longer limited to one single substrate (digital materials), and if the set of available materials (digital, physical, and smart) is growing at a rapid pace, then the biggest challenge is not finding ways to manifest interaction in material form; instead, the challenge is navigating this landscape of available materials and devising a method and an approach for doing this job. In this section I will present one such approach to material-centered interaction design, which I have labeled the "interaction-first" principle.

If we have lots of different materials available, and we have good knowledge about how these different materials can be combined and integrated, then we can feel confident that we will find an appropriate material form for the interaction we imagine. However, imagining the interaction needs to come first, and then it can and will be refined through the process of trying to manifest it in material form. The implication for this reasoning is straightforward: the idea of the interaction—its form, function, meaning, and way of presenting itself—needs to be expressed in order for the interaction to then be explored through a material lens.[16] This is what the interaction-first principle is all about.

Let's take driving as an example. Driving (a car) is not only an activity, or somehow related to a means of transportation (going by car), but it also says something about the relation between the user (of the car) and the role of the user in relation to the car during transportation (driving demands an active user throughout the ride). In short, the user becomes a user through driving as an act of interacting with the car. Driving is here the "mode of interaction," and given that we design for this particular form of engaged mode of interaction, a self-driving car is not the solution here (and self-driving cannot manifest the interaction model of the user driving the car). However, if the activity is rephrased as "a relaxed mode of transportation," then the interaction model is different, the assumed relation between the user and the car is different, and it will call for and even open up alternative design solutions (now the self-driving car is suddenly a possible design option). So the mode of interaction is crucial for how the design space is framed. When a designer has no idea or understanding of the form or mode of interaction, any attempt to manifest interaction in material form will likely fail. For this reason, interaction design projects need to start with an idea of the intended mode and form of interaction first—hence, the "interaction-first" principle.

The interaction-first principle is about conceptually defining the mode and form of the interaction being supported. It is about defining who the user is and how he or she will interact with the interactive system.[17]

What might a method for doing material interaction design look like if it followed the interaction-first principle? I will sketch out some basic steps of one such process here, and then illustrate how these steps unfold with a practical example.

A method for doing material-centered interaction design following the interaction-first principle includes:

(1) Exploring and defining the form of interaction being designed,
(2) Exploring and evaluating the range of possible materials that can be used for designing the interaction,
(3) Working iteratively between the integration of different materials in the design of the interaction, while recurrently revisiting and if necessary revising the initial idea of the interaction.

As we see here, the focus is on how a particular form of interaction can be manifested in material form, but each step of this simple method is also about constantly keeping a keen eye on the idea of the interaction being designed. As such, this is a multilayered iterative process—including not only iterations on the level of the materials in terms of "having conversations with the materials at hand," or in terms of needing to revisit and reevaluate different material choices and design solutions. In parallel to that, it is also a process of iteratively revisiting the core idea—that is, the idea of what form of interaction is being designed.

If we take a close look at the different steps involved here, we should notice that the first step involves not only exploring and defining the form of interaction and how it should unfold between the user and the artifact, but also considerations related to the context, surrounding, or environment in which the interaction will take place. Further, step two—exploration and evaluation of different materials to be used in the design—is a process that relies on skills and expertise in understanding this range of materials and experience in working with them. The understanding and knowledge of materials includes a good understanding of material properties, as well as how different materials react to or work in concert with other materials. And finally, the third step of this method demands skills for working on a concrete level with the materials at hand, while simultaneously working on a more conceptual level. The latter involves constantly evaluating whether the interaction concept can be implemented with the current materials at hand and/or whether the interaction concept needs to be developed or altered, given how well the materials will suit the design in its material instantiation.

If we take one step back, we can see how well the interaction-first principle works in relation to a material-centered approach to interaction design. This principle focuses on the form of interaction being designed, and it gives the process a conceptual focus on what the aim of the interaction design is. This conceptual focus helps as a conceptual guide throughout the material-centered design process, and it helps to explain the interactive artifact after it has been designed. From a material-centered perspective, the interaction-first principle also works as conceptual benchmarking tool, a conceptual ground for the practical work of crafting, programming, and polishing the interactive artifact being designed. In short, the interaction-first principle suggests that interaction design needs to start with a clear idea of the inter-action, and then dress it up in terms of its material instantiation.

One can think that this leads to an interaction design process that is solely conceptual and material-centered. However, this approach does not stand in contrast to any user-centered approach to interaction design. On the contrary, the "interaction-first" principle is all about understanding user needs and about imagining good, interesting, and engaging interac-tion design solutions in relation to those needs.

Accordingly, if the "interaction-first" principle is well executed, it will to a large extent be about user experiences, and about how the user will engage in the act of interaction. With that conceptualization in mind, the interaction designer has the task of meeting those expectations in the tran-sitional process of crafting those ideas in material form.

For an example illustrating the "interaction-first" principle and how it guides a material-centered approach to interaction design, let us look at the design process for the Apple Watch Series 2. Included in this process was a design challenge the designers faced in relation to the mode of interac-tion imagined, and their reconfiguration of the material composition of the watch to enable this particular mode of interaction.

When Apple launched the new Apple Watch in 2016, their design direc-tor Jonathan Ive explained that they had faced a particular design chal-lenge during the design and development of this watch. Although the first version of the watch had worked well, it was not waterproof. To enable the watch to be used in any context, even when swimming or surfing, the design team not only had to make the watch waterproof, but had to do so without taking away the built-in speaker (which would then make the watch less useful for reading messages, playing music, and so on). The clas-sic problem that speakers have with water is that they require a vibrating membrane that produces sound waves, so if the membrane is blocked with water, no vibrations can make these sounds. As a result, the combination of the two materials, electronics and water, is typically not a good idea. The design challenge was thus not to keep these two substrates apart, but rather to find a way for them to work in concert with the speaker membrane, as yet another design element in this equation.

To address this design problem, the team started to explore alterna-tive design solutions that would (1) enable a waterproof smart watch, while also (2) arriving at a design with a built-in speaker. The solution the Apple

Figure 4.4
An illustration of the differences in material configurations between the first version of the Apple Watch (left) and the Apple Watch Series 2 (right).

Figure 4.5
The user interface developed for the Apple Watch Series 2 to enable a user to unlock the watch and eject water from the watch's interior after it has been used in water.

design team came up with was the unique design of the Apple Watch Series 2, which enables a waterproof smart watch to have a working speaker and a user interface that its user can utilize to eject the water from inside the watch after it has been worn in water (for instance, during swimming). Figure 4.4 illustrates the material reconfigurations that had to be made to the first version of the Apple Watch to make the Series 2 watch waterproof, and figure 4.5 illustrates the user interface developed to let its user eject the water from inside the watch after it has been used in water.[18]

With this design solution, Apple managed to address the design challenges posed by using the watch in the water. They achieved this solution by using the speaker not only to produce sounds but also to use the same technique—the creation of vibrations—for the purpose of ejecting water from the inside of the watch. From the perspective of a material-centered approach to interaction design, this design demonstrates that new and novel design solutions can result from a good understanding of materials and how materials work in concert (in this case, how the speaker membrane can be used to produce vibrations that cause the water to come out of the watch). Understanding how water (as a material) reacts to vibrations (as another material) was key for arriving at this particular design solution.

On a more general level, this example also illustrates how the design team managed a design process that was about the simultaneous development of (1) a clear interaction model, while focusing on (2) the manifestation of this model in material form. In doing so, this example suggests what the process of "form giving" and "form making" could look like in the context of interaction design.

To summarize, this chapter introduced the "interaction-first" principle and discussed how the "form of interaction" is designed through a design process where we move from a conceptualization of the interaction to the material instantiation of that interaction.

So how can we design interaction across different substrates? The Apple Watch worked as an illustrative example, but what can be said on a more general level? And what are the properties of interactivity that one can think about when imagining and conceptualizing "the form of interaction"? And finally, how can that form of interaction be brought into composition with materials? In short, what are the principles for working with part-whole relations in interaction design? Those are the questions I will address in the next chapter, on interaction design across substrates.

5 Designing across Substrates
On Interaction Design with and across a Wide Range of Materials

Is interaction design a *single-material* design tradition? Just as other types of craftsmanship have focused on a single material (including knitting, woodcraft, or ironwork), we can ask whether interaction design as an area of craftsmanship can be said to focus ondigital materials as the "single material" of interaction design. At first glance, it is tempting to say yes; if we review the total outcome of interaction design projects, we can probably say that 99.9% of all interaction design projects are screen-based, and maybe even web-based. Furthermore, all of these projects are made out of code, and even the tools, including the programming languages, scripts, and code compilers, are made of digital materials (ultimately, bits).

Early work with punch cards and the design of the computer mouse by Douglas Engelbart demonstrate a different interaction design tradition, however. The punch cards were made of paper, and the mouse was partly made out of a piece of wood. The punch cards worked in combination with electronics designed to read the cards, and the wood used in the design of the computer mouse not only functioned as a container for the electronics inside, but manifested a particular form of interaction in the form of a hand-computer interaction modality. These two examples illustrate the typical necessity for many interaction design projects—to integrate different materials in one design. And again, if taking a contemporary example like self-driving cars, it becomes obvious that a self-driving car is not possible without an inseparable integration of physical and digital materials. It is the deliberate composition of these substrates that enabled the self-driving car to move from a concept to a part of our reality.

In this chapter, I take as a point of departure a material-centered approach to interaction design. In particular, I investigate whether contemporary interaction design is restrained to only one material, as I have started to ask here. Before going further, I would like to say that the answer to this question is no. A material-centered approach to interaction design cannot only be about digital materials. On the contrary, contemporary interaction design is about the configuration of multiple materials to work in concert. Interaction design thus becomes a practice of designing across a multitude of substrates.

On materials and six different dimensions of interactivity

What does it mean to say that something is interactive? In order to do good interaction design, it seems important to have an idea of what *interaction* is, what *interactivity* is, and what it means if we say that something is interactive. Otherwise, a blurry understanding of the object of design (in this case, "interaction") might lead to equally blurry design solutions. In this book, the answer to this overall question is quite straightforward: "Interaction" can be defined as *inter-actions between two or more entities,* and so "interaction design" is the design of these interactions in relation to the entities taken into account during the design process. As I will show in chapter 7, these interactions in the form of "threads of interaction" together with "threads of processing" performed by the computer co-produce the materiality of interaction.

"Interactions" means actions in between two or more entities; the entities can be a user and a computer, but they could also be any other sets of entities. For instance, the concept and realization of a self-driving car involves a complex set of interactions in between quite a large set of different entities that together constitute "the materiality of interaction" in the context of a self-driving car. There are, most importantly, the two entities of the passengers and the car, but there can also be other cars, pedestrians, bicycles, and even wild animals to take into account in the design of a self-driving car—just to mention a few sets of entities. Furthermore, a self-driving car must be designed in relation to traffic rules as yet another, although maybe more abstract, entity, and in relation to other interactive systems, including the GPS system. So, even though a passenger in a self-driving car might be imagined as not interacting that much with the car as an interactive entity, the car as an interactive entity is interacting, on behalf of its passenger, with lots of other entities, and not only in machine-to-machine interaction (for instance, with other self-driving cars).[1]

This example illustrates the "relational character" of an interactive system. In addition, it illustrates how interactivity permits, and enables, something static (like a physical car) to become or behave dynamically in relation to other computational or noncomputational entities. This dynamic is what we see as a result of "inter-actions" between different entities.

Löwgren and Stolterman (2004) have similarly described interactivity in terms of a digital artifact's dynamic gestalt. And according to Janlert and Stolterman (1997, 138), "the dynamic gestalt of a digital artifact is, in this sense, rather like its overall character." Further on, the word "dynamic" suggests that the artifact is interactive, and so it is only through the use of the artifact that we can fully understand its overall character. In an attempt to further explore the dynamic character of interactive artifacts, Lim et al. (2007) analyzed the notion of the gestalt both in relation to interaction and in relation to aesthetics.

So this notion of "dynamics" or "dynamic gestalt" describes a digital artifact in terms of how its appearance and behavior change during its use. From an interaction design perspective, this "dynamics"—this ability to change through use and to communicate a change of state to its user—can be deliberately designed. This deliberate design of changes of state is what interaction design is about, to a large extent. Good interaction design allows its users to see and understand that the artifact is dynamic, and the easiest way to communicate this is to enable the artifact to clearly communicate its change of state and options for further interaction.[2] In a sense, then, the "dynamics" of an interactive artifact serves two purposes: it is what makes the interaction possible (in that the artifact responds to operations), and at the same time it moves the interaction forward, in that, through its different states, it tells the user about its current state and available options for further interaction. It is this dynamic form that needs to be designed when doing interaction design, and in order to do that well, it is important to have a good understanding of the different dimensions or properties of interactivity that come into play when designing interaction. In this chapter I will therefore sketch out six such aspects or properties of interactivity.

Properties of interactivity

Digital artifacts can manifest interaction in many different ways. When I say "manifest," I am referring to how a digital artifact is given its materiality of interaction. If we think of this process as a *postrepresentational paradigm of interaction design*, the implication is that computing is no longer only about *representing reality* (as it follows from the *representation-driven paradigm*), but it is actually *part of reality* (and in that sense "postrepresentational"). Computing has become an inseparable part of our everyday lives, environments, and activities, and accordingly it is no longer only about representing some aspects of this reality. Far beyond that, computing is part of our reality.

At the same time that computing is blending increasingly with our everyday reality, it is also increasingly disappearing.[3] We can notice this not least on a literal level, as computing hardware is increasingly miniaturized and embedded and integrated in physical objects, vehicles, and buildings. Today we might not even visually recognize that a physical object also has embedded computational power. The seam between the physical and the digital worlds is increasingly dissolved.

However, from the viewpoint of the materiality of interaction, no matter how embedded or invisible digital technology is, it still displays itself in a wide variety of ways to enable interaction with its users. So, from that particular perspective, we need (1) a good understanding of different dimensions of interactivity, in order to (2) enable a discussion on how these dimensions can be manifested in interaction design. In short, it is important to understand the design space for the materiality of interaction, which is ultimately restricted by our understanding of the different dimensions of interactivity.

To understand what I mean when I say that there are many different dimensions of interactivity, think back to the beginning of chapter 3, where I discussed why we cannot just say that interaction design is about the design of user interfaces,[4] or the design of the turn-taking back and forth between the user and the computer. When I say that there are many dimensions of interactivity, I mean that it is exactly along these dimensions that interaction is designed (implicitly or explicitly), and therefore, a good understanding of these dimensions helps us to understand the design space of interaction design—or, put slightly differently, the form factors of interactivity.

But what are the dimensions of interactivity I have in mind here? What are the form factors of interaction that defines this design space? If we think about interaction with a focus on the "in-between" and with a focus on how interaction plays out as an *inter-act* between two or more entities, as I discussed in the beginning of this chapter, these six dimensions include: (1) *changes of state (dynamics)* in the interaction, (2) speed of change (pace), (3) requests for input (turn-taking), (4) *responsiveness to input (receptiveness)*, (5) *single-threaded or multithreaded interaction,* and (6) *direct or agent-based ("invisible") interaction.*

In what follows, I will detail these six dimensions of interactivity and give some examples of how these dimensions manifest themselves in material form in some typical interactive systems.

(1) Changes of state (dynamics)

Interaction with computers, even on a micro level of interaction, is very much about achieving a particular change of state. Even when Douglas Engelbart demonstrated the computer mouse for the first time, we can notice this interest in seeing interaction as a matter of changing a current state into a preferred new state through the act of interaction. As part of this classic demonstration, Engelbart placed the mouse pointer at a particular location on the screen and said, "Let's add some materials here." He moved the mouse pointer to that particular location and typed in a couple of letters. As he moved the mouse pointer over the screen, he changed the state of the screen and showed an animation of a moving mouse pointer; and when he typed in a few letters, he again changed the state. With the added material on the screen, he opened up space for the next stage of that particular interaction. For instance, adding some text on the screen allowed for further interaction such as marking up, editing, or deleting that particular piece of text.

Beyond this micro level of interaction, we can notice how this dimension of *changes of state* scales to the level of interactive artifacts, systems, platforms, and websites. For instance, interaction is typically built up around a series of different screens. Site maps illustrate this for larger websites, and the whole practice surrounding wireframes and tools for screen layouts illustrates the need to carefully plan how interaction should unfold as a sequence through a number of different stages, screens, interfaces, or dialogue boxes.

As a dimension of interaction design, changes of state are the way the design enables interaction that takes the user from one stage to the next one. On a website this might be slow, and to a large extent user-initiated (when the user clicks on a link or a button), but changes of state can also be a dimension in much more dynamic designs. For instance, in computer games, user-initiated changes are not the only thing that affects this movement from one stage to the next. Instead, it is the combination of user-initiated actions and counteractions from the game that typically makes arcade games interesting and challenging. For instance, typical arcade-type computer games use this dynamic to present challenges to the gamer (for instance, to jump over obstacles appearing on the screen). The game layout changes and the challenge is to a large extent about timing (to jump, turn, run, or shoot at exactly the right moment). As such, this dimension of changes of state is in many cases related to the next dimension discussed here, which is the *speed of change*, or the *pace* of the interaction.

(2) Speed of change (pace)

Besides *change of state*, a related dimension of any interaction design is the *speed of change* (and accordingly the speed of interaction). What we can do (or change) in different dimensions of interaction design is, of course, important, but so is the pace at which we can interact. As a design variable, *speed of change*, or the *pace* of the interaction, is something that can be designed, yet unfortunately it is easily forgotten in interaction design projects.

In computer game design, pace is a central design dimension that is constantly foregrounded. In many games, in fact, the pace of the interaction required to solve a puzzle, or complete a level, is of essence. The higher the level, the quicker the interaction needs to be. Typically, that is how to make a computer game more challenging for each level.

As an interaction design dimension, we can think about pace in terms of quick versus slow modes of interaction. Compare computer games to the checkout procedure on most web shops, which can be seen as an example at the other end of this scale. In computer games, fast reactions, and thus fast interaction with high precision, such as quick hand-eye coordination, are often needed, whereas in the design of traditional web shops, the interaction through the checkout procedure from the web shop is designed to be slow enough so that the user will enter the right payment and delivery details, and will feel secure at all points during the process. Again, pace is a dimension of interaction design that needs to be deliberately designed.

Another example is the interaction design of information kiosks. Typically, these are designed as guides to help people find information. It is assumed that people approaching an information kiosk lack information and need help. A typical interaction design for an information kiosk thus builds around a slow-paced interaction model. It is assumed that the user might need some time to understand what to enter into the system and also that the system should present information and alternatives for interaction at a slow pace. This slow speed is intended to prevent confusing a user who is already lost to some extent (in terms of lacking some needed information).

Even though a slow pace of interaction might be the most common interaction design for information kiosks, this does not mean that the design should rule out opportunities for faster interaction. Yet this option is often a forgotten design dimension of many interaction design projects. For instance, many information kiosks do not allow users to type their input quickly, even though a user might know exactly what to type. Instead, that part of the interaction (user input) is also designed to be slow, forcing the user to type numbers or letters very slowly so that the computer can register every keystroke. While this might be just an overlooked detail, it is again a dimension of interaction design—from the viewpoint of how the materiality is structured to allow for a particular form of interaction.

(3) Requests for input (turn-taking)

To design interaction also entails designing the balance between what the computer does and what the user does. Thinking about interaction as a turn-taking act between human and machine, we can ask: What should the balance between these two entities look like? Can some interaction design projects benefit from one particular balance and other interaction designs benefit from a completely different balance between human and machine?

The classic command line interfaces of the 1970s and '80s (before the arrival of the graphical user interface) were a balanced interaction model in that the computer and its user did equally much to make the interaction move forward. The computer presented some options, the user entered a command, the computer responded to the command, and then the user typed in the next command. As such, the interaction model was a simple

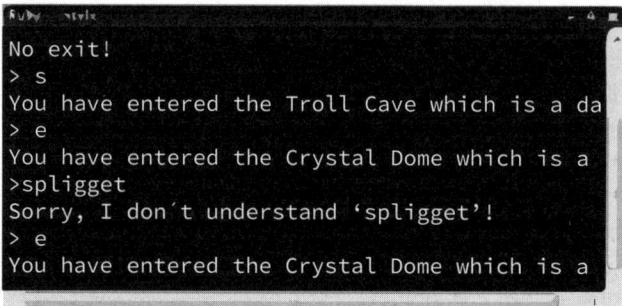

Figure 5.1
Example of balanced interaction model for total symmetry in the turn-taking between a user and the computer.

turn-taking act back and forth between a person and the machine. In fact, during the 1970s, 1980s, and even part of the 1990s, some computer games were not only text-based but also followed this simple, linear form of interaction.

Can we think of alternative configurations of the interaction model, other than complete symmetry between human and machine, in the distribution of the interaction workload between these two entities? There are certainly a full scale of alternatives here—from user-driven models of interaction to computer-driven models of interaction.

A couple of simple examples can further explain these different alternatives. These examples illustrate different balances between user and machine from the viewpoint of turn-taking. The focus is on the interaction, which might not necessarily correspond directly to either the cognitive load for the user or the CPU power needed to process a particular set of instructions.

In the example of user-driven interaction, most of the interaction and, accordingly, the source for changing the current state of the interaction depend on what the user does—and not on additional elements added by the computer. The traditional word processor is a good example. When a user types, the word processor reflects only the user input, by representing each keystroke on the screen. Of course, this is an oversimplified example, as modern word processors are capable of doing much more sophisticated things, but this simplest form of using a word processor exemplifies what "user-driven interaction" means in this context.

On the other side of this scale is computer-driven interaction. Here we have the opposite division of workload between user and machine, where the machine is doing most of the work and the user only now and then interacts actively with the computer. An example is the use of computers for media consumption. In the typical use scenario, when a user is listening to music via a computer, the machine plays each song, follows a particular playlist, streams music from a server, and maybe also distributes the music played to wireless speakers. The computer might be remotely controlled via another device or app, and the user might only need to interact with the computer to switch to a different song or playlist, or adjust the volume. With the computer doing most of the work, the interaction is configured to allow the user to relax or to do something else while listening to the music.

These two examples illustrate two alternative ways of distributing the interaction workload between user and machine. As an interaction design dimension, the input balance can be designed for an interaction that involves lots of requests for input from the user or designed to be the complete opposite, with as little interaction demanded from the user as possible.

(4) Responsiveness to input (receptiveness)

When we think about something being "interactive," what comes to mind is a digital artifact that dynamically reacts to user input. In this case, the interactivity of the digital artifact is defined by its *responsiveness*—that is, the extent to which it responds to user input. Responsiveness also occurs on a scale ranging from fully responsive to nonresponsive. If we think about how the level of responsiveness can shift during sessions of interaction, we can actually and deliberately design this level of responsiveness. Responsiveness to input—or the extent to which a digital artifact is receptive to input at a given moment—becomes a design dimension for interaction design.

As an example, during runtime, a computer program executes lines of code. This code is typically designed to call different procedures, access databases, run loops, use timers, process information, and wait for user input. However, from an interaction design perspective, there are many opportunities here for how these and other aspects of a computer program are combined, and thus how the program behaves during runtime. For instance, does the program allow for simultaneous activities to be executed during runtime, so that although the computer is processing some user input, it can still receive other input as well? Most web forms do not allow for such interaction, but many computer games certainly do have such solutions. In fact, in many computer games it is a necessity that the game not "freeze" until the computer has processed some user input. Interactions did "freeze" in the early days of computing due to the limited processing power, but that is not the case anymore with contemporary computers. Instead, the level of responsiveness is something that can be decided upon and designed.

(5) Single-threaded or multithreaded interaction

Now that the level of responsiveness to input is no longer dependent on computing power and today's operating systems allow for multitasking, so that different programs can share the same hardware and CPU power, interaction design can range from single-threaded to multithreaded interaction.

Single-threaded interaction is characterized as linear and step-by-step interaction. For instance, when we click through a web form, this is the typical model for the interaction. In single-threaded interaction, the digital material is organized so that there is one clear "path" through the software, typically arranged as steps, and the software provides no or little support for breaking away from this path of interaction. Typical examples include self-check-in machines at airports or ticket machines for train tickets. Another classic example is the traditional analog telephone. The classic phone was very much about one standardized thread of interaction, which included the fundamental steps of picking up the phone, waiting for the dial tone, dialing the number, waiting for someone to answer, having the conversation, and finally hanging up the phone. This design constrained human-phone interaction to unfold in this particular order, and it was very hard to break this particular single-threaded path of interaction.

At the other end of this scale we have the modern smartphone (or just about any general-purpose computer, like a laptop). With multithreaded modes for interaction available in contemporary smartphone designs, the human-phone interaction model is no longer restrained to one particular path. Instead, the user can choose to do different things, and in an almost infinite number of orders, including cutting and pasting information from one application to another, moving seamlessly between several applications during runtime, going back or undoing commands, or waiting until later to continue a particular task. This mode of interaction does not provide challenges related to the distribution of processing power, or how to design operating systems that allow for multitasking, but it is challenging in that the designer must imagine the different paths users might take across the digital services being designed. In short, this dimension which enables multithreaded interaction models simultaneously opens up new interaction design challenges.

(6) Direct or agent-based ("invisible") interaction

A final interaction design dimension is the user's level of involvement *during* sessions of interaction. In short, is the user directly involved, or for the most part indirectly involved via the actions taken by an agent?[5]

As an example, it makes sense to design for direct interaction with a word processor, but in the case of self-driving cars, maybe we want to design the user experience so that the user can just say, "Take me to point B," and then the rest of the user's interaction with the car has to do with whether he or she experiences the ride in the car as safe, efficient, and so on. In the example of the self-driving car, we design the interaction with a point of departure in an interaction model that assumes very little direct interaction, and we rely on an agent-based interaction during the actual ride in such a car. With the growing number of human-robot models of interaction, this agent-based invisible mode of interaction is something that will grow as a perspective when designing interaction.

One might think that the advent of the Internet of Things (IoT), which includes the full range of classic examples like refrigerators that tell their owners they need to buy milk, robot vacuum cleaners, or automatic lawn mowers, is the trend that is motivating our thinking about interaction design in terms of direct modes of interaction versus agent-based modes of interaction. However, this debate is in fact a classic discussion in the field of human-computer interaction. For many years the field has had a continuous discussion on the pros and cons of direct manipulation versus interface agents. The classic paper on this topic, Schneiderman and Maes (1997), discusses these two very different modes of interaction.

From an interaction design point of view, implementing interactive systems that rely on direct manipulation is indeed very different from designing for agent-based interaction. When designing for direct manipulation, the "directness" in the interaction is crucial. This means that the user needs to have direct access to what the interface offers, including not only functional access, but a self-explanatory interface so that the user understands how to interact with it at any given moment. This aspect is crucial for the "directness" in the interaction to work. In the 1980s and '90s this sparked a whole stream of discussions in our field on interface "affordances" and interface/interactive systems usability.[6]

On the other hand, agent-based interaction is all about agency.[7] The crucial thing here from an interaction design perspective is not the ongoing direct conversation or constant turn-taking with the system, but ensuring that the system does what it is asked to do and reconnects with the user once the job is done or when additional interaction is needed. Overall, this means that direct manipulation allows for tight coupling between user actions and corresponding representations of those actions in the computer system, whereas agent-based interaction allows for less frequent or loose coupling between what the user wants to accomplish and how the computer goes about accomplishing that task. This might range from how a search engine finds some information on the web to how a self-driving car navigates traffic during rush hour in the city to take its passengers to a particular destination. In agent-based interaction, what matters is what the computer accomplishes based on the instructions provided by the user.

Based on these two extremes—direct manipulation versus agent-based interaction—it is obvious that this is also a scale that needs to be taken into consideration when designing interaction.

While agent-based interaction was a vision for many years, it is now rapidly being established as a common mode of human-computer interaction. In fact, more and more interaction designers rely on sensors to collect data or to retrieve input. As an example, the app Runkeeper relies on GPS data to track movement. (This is a good alternative from an interaction design viewpoint. Just imagine a "direct interaction" alternative in which the user would have to manually enter every new location either by clicking on a map or punching in longitude and latitude while running.) Automating user input by using sensor data or scripts can certainly be beneficial in many use cases.

This trend of moving beyond direct manipulation as the main mode of interaction is also pushed forward by a number of technological trends at the given moment. For instance, it is now very easy for users to create scripts to automate interaction.[8] Advanced sensor technologies and algorithms are further examples of technologies that enable flows of information and information processing to be partly automated. There are also a few technologies that enable automation for otherwise manual tasks: when we synchronize our devices, either between devices or between devices and cloud-based services, this concept of "synchronization" demands only an initiation by the user and then the whole process is automated to a great extent. And in some cases, even the initiation follows a certain script or rule, such as "Synchronize the device when it is online over WiFi" or "Search for new firmware updates every day at 07.00 AM."

Finally, the concept of "batch processing" is yet another associated technique that allows for computing to happen while we're busy with other activities. Batch processing is the execution of a series of jobs in a program on a computer without manual intervention. As such, true batch processing is noninteractive from a user perspective.

How can we put this type of materiality into the context of material-centered interaction design as presented in this book? The common quality shared by these scripts, sensors, algorithms, synchronizations, and batch processes is that they can be configured to execute a series of program instructions as part of the interaction design. The ordering of program instructions into such executable series is typically referred to as "scripts," and so I suggest that in order to do interaction design through a material lens, we need to think about interaction design with such *scripted materialities*.

"Scripted materialities" typically defines both the temporality of the interaction (that is, the *pace*) and the balance between the user and the computer in terms of the design of the *turn-takings* needed in the interaction. This notion of "scripted materialities" also suggests that the material is not only dynamic (in that it can change its current state) but can also be programmed to automatically execute commands or let the computer process code and information. In short, any interaction designer who takes on a material-centered approach to interaction design also needs to consider how the materiality of interaction can be partly scripted, and how such scripted parts of the interaction will affect the overall experience of the interaction.

These different design dimensions illustrate that interaction design is not only about user interface design (although that is, of course, also important), but also about the design of the interaction and how it is intended to unfold. If we see interaction design as this process of forming interaction—by carefully designing its pace, states, and possibilities for change, whether it allows for multiple interactions or is a single-track interaction model, and so on—then we shift focus from seeing only user interfaces as the materiality of interaction toward acknowledging how all these design decisions are ultimately manifested in computational material form.

Once we look at it this way, it becomes obvious that good interaction design demands (1) a good understanding of the dimensions at play when designing interaction, (2) a good understanding of how different materials can be formed and combined to scaffold a particular form of interaction, and (3) a skilled mindset for imagining particular forms of interaction and how one such form can be manifested. In short, we see how a particular idea of interaction can be given an appropriate materiality.

It is important to know what kind of form of interaction is being imagined in order to manifest it in material form—to give the interaction a proper materiality.

Activating and combining materials in interaction design

The notion of "scripted materialities" highlights how materials can be programmed to execute series of commands as part of an interactive system. In chapter 4, I described how contemporary interaction design in many cases stretches across many different materials. But how does this actually happen? What are the opportunities and techniques available for activating and combining materials in interaction design projects?

In this section I will discuss how materials might be activated via *sensors* and *actuators,* and will talk about *linkage* in terms of how materials might be combined in interaction design projects.

Sensors and actuators

Sensors enable computers to "sense" and collect data about different aspects of our world. For instance, photodiodes can collect data about light conditions, accelerometers can detect rotation, pressure sensors can detect forces, and so on. When a computer has access to connected sensors, it can "read" its environment, and so the environment or other materials, things, temperature, locations, and so on can function as input to what the computer is processing.[9] But how can computers also "write" or affect their surrounding environment? This is where actuators come into the picture.

Typical Sensor & its Output

Property	Sensor	Active/Passive	Output
Temperature	Thermocouple	Passive	Voltage
	Silicone	Active	Voltage/Current
	RTD	Active	Resistance
	Thermistor	Active	Resistance
Force/Pressure	Strain Gage	Active	Resistance
	Piezoelectric	Passive	Voltage
Acceleration	Accelerometer	Active	Capacitance
Position	LVDT	Active	AC Voltage
Light Inensity	Photodiode	Passive	Current

Figure 5.2
A schematic overview of some sensor technologies.

Actuators enable computers to move or in other ways affect other items and materials in their surroundings. An actuator can, for instance, be a servomotor that can rotate its axis to a particular position. However, the opportunities can really be explored when a combination of sensors and actuators with a controller (a processing unit) is arranged as part of some form of human-computer interaction. For instance, combining a sensor that reads the current outdoor temperature with a servomotor can enable a computer to close a building's open windows if the temperature drops below a preset value, or can send a notification to a user together with an option to close the windows. This is only a simple example, but it illustrates how sensors and actuators enable computing to be integrated with aspects of our surroundings on a very practical level in terms of material and functional integration.

There are many different types of sensors available today. Some of these are active sensors and others are passive; each can sense different aspects of the environment (for instance, light, pressure, or position). The output from these sensors can also come in different forms, such as in the form of voltage, current, or resistance. For a schematic overview of some sensors, see figure 5.2.

Furthermore, sensors, controllers, and actuators, together with options for users to interact with such configurations, enable interaction designs in which loops circling sensors and actuators are deliberately designed. For example, a sensor can read a value and an actuator can be programmed to

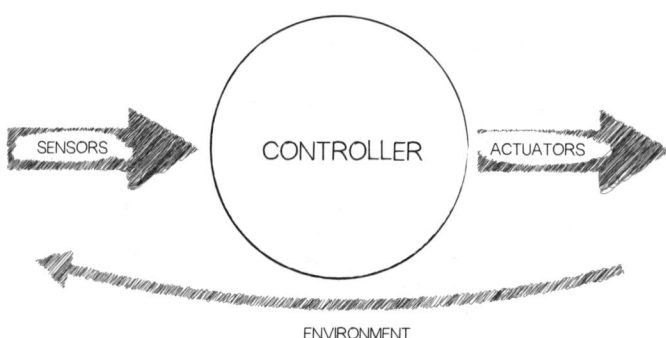

Figure 5.3
Model illustrating the sensors-controller-actuators-environment cycle.

take action given a certain sensor value. This action might further affect the environment, and the sensor can read this change as input for further processing, and so on. A typical example here is a thermostat that almost by its definition implements this model in terms of sensing an aspect of the environment (temperature) and then regulates a heater in relation to the values read by its sensor. For a more schematic overview of this arrangement, see the illustration in figure 5.3.

There are several different types of actuators; the two most common ones are electric actuators and hydraulic actuators. Typically, an actuator takes the form of a motor that enables a computer to move, rotate, or change the position of something in the physical world. Electric actuators can be categorized as DC servomotors, AC servomotors, stepper motors, and solenoids, whereas hydraulic actuators use hydraulic fluid to amplify the controlled command signal in order to perform push or pull actions.

With actuators in place, a computer can execute program instructions that have physical consequences for other materials. For instance, a servomotor can be configured to bend a piece of wood, or configured in other ways to affect other materials.

Overall, we can say that actuators enable the computer to "reach out" to affect or activate other physical materials. While sensors enable the computer to sense and "read" its environment, the actuators enable the computer to also take physical actions—to "write" in material forms. In the area of robotics, this has been known for many years and is a fundamental principle behind robotic design. However, for interaction design, this clearly opens up opportunities for working not only with digital materials but also across the full range of imaginable material and immaterial substrates in interaction design projects.

Linkage (and connections)

The use of sensors and actuators illustrates one way to activate and combine materials in interaction design projects, but there are also other techniques available. In this section I will talk about a number of such technologies under the label of "linkage." I chose this label because these technologies are all about linking different parts together into larger computational arrangements. There are, of course, many technologies available for linking different computational parts together, so in this section I will go through only a small number of those to illustrate how linkage is possible. These different technologies include functions, remote procedure calls (RPCs), object-oriented programming, protocols for data transfer and computer networking, linked open data, service-oriented architectures (SOAs), and server query languages.

In computer programs, functions are snippets of code that can be called from other parts of the program. The program can send one value or several to a function and receive a value back from a function. The use of functions enables the design of computer programs beyond a completely linear structure. In nonlinear programming, we have both the early solutions of RPCs and more recent object-oriented programming. Let us first take a look at RPCs.

In distributed computing, an RPC occurs when a computer program causes a procedure (subroutine) to execute in another address space (commonly on another computer on a shared network), which is coded as if it were a normal (local) procedure call, without the programmer explicitly coding the details for the remote interaction. That is, the programmer writes essentially the same code whether the subroutine is local to the executing program or remote. With RPCs, computing can easily be run across several computers during runtime.

Another example of a technique for linkage is the whole philosophy underpinning object-oriented programming. The idea of object-orientation arose from a need to move away from linear programming, to break up long chunks of code into small independent parts and then to formulate a framework so that the parts, or objects, could be integrated and work together.

Computing happens not only across objects but also across multiple computers. From the early attempts to do distributed computing, via the development of GRID networks, to the cloud services of today, the trend is clear—computing is increasingly done across many connected computers. These computers need to be able to communicate and to exchange data. For that purpose, a number of different protocols have been developed. Again, these protocols are examples of technologies that enable different computational materials to be linked together.

There have been many recent developments in the linkage of data. For instance, the development toward "linked open data" adds design opportunities to the web to connect related data that wasn't previously linked, or to use the web to lower the barriers to linking data currently linked using other methods. More specifically, one can say that "linked data" is a term used to describe a recommended best practice for exposing, sharing, and connecting pieces of data, information, and knowledge, and in more technical terms it is usually described as part of the Semantic Web (and the use of URIs and RDF).[10]

While linked open data is a quite recent movement, other frameworks for data integration have been around for many years, such as SOAs, a framework developed to support integration of different systems. An SOA is a type of computer software where services are provided to the other components by application components, through a communication protocol over a network. As such, SOAs illustrate not only how different parts or "components" of a system can be integrated, but also the use of protocols and computer networks to enable this type of linkage.

Finally, database technologies, including server query languages, might be one of the most commonly used technologies for storing, processing, and accessing data. With databases, enormous amounts of data can be collected, stored, shared, and processed through queries. As such, databases also enable data to be an integrated part of interactive systems design.

Interaction design—Across substrates

Interaction design is about the design of user interfaces—whether the interface is screen-based (for instance, graphical) or screenless (for instance, voice-based, gesture-based, or tangible).[11] As the beginning of this chapter suggested, there are a number of additional dimensions to take into account when designing *the form of interaction*—including the design of the pace of interaction, whether it should be a single- or multithreaded mode of interaction, and so on.

To design a particular form of interaction, many different computational resources might need to be integrated, and there might be a need to work across many different materials and substrates in order to enable a particular form of interaction.

Today, interaction design is not only manifested in digital materials (like code and scripts) or provided via computing hardware (such as in the form of graphical presentations via high-resolution screens). Instead, computing takes many forms and interaction happens across many materials, devices, and other objects. For example, if we go running with Runkeeper on our smartphone, the smartphone is not the only digital device that is part of the interaction. Far beyond that, the shoes we wear, the path we decide to run, and the GPS satellites in orbit around the earth are part of the interaction that monitors the run. The run is not only about exercising, but also works as the main input to the Runkeeper app.

The lesson to be learned here for interaction designers is that interaction design is more than just about designing an app or any other interactive piece of technology. In addition, it is not limited to the design of this technology in relation to an activity (in this case, the activity of "running"). Beyond this, it also includes other materials, technologies, and objects. Today, interaction design is to a large extent about figuring out how all of these things can be combined in meaningful and usable ways (and in some cases even thoughtful and/or entertaining ways).

Indeed, it is more of a rule than an exception nowadays that interaction design is about the design of sessions of interactivity that might "live" across many different devices, operating systems, platforms, and services. For instance, the Internet bank needs to be accessible from a stationary computer, but also from a laptop, tablet, smartphone, and smart watch. With ongoing sessions over many different computational resources, then, interaction design is no longer a single-device design practice, but on the contrary is a "relational practice" that is heavily focused on combining and uniting different materials, devices, and services so as to enable streams of what I call *inter-actions*.

This trend is likely to continue. Just think about how we might extend this current practice to further combine different materials, actuators, sensors, and input/output modalities in the design of new interactive artifacts and digital services. Not only might there be more complex interaction landscapes evolving in terms of the combination and integration of these different parts into larger arrangements and ongoing sessions across different devices and services, but they will also be more complex, in that services and sessions will also live across switches between active and more passive modes of interaction.

As an example, think about the mode of interaction while playing a computer game in comparison with the mode of interaction while wearing a personal tracking device like a Fitbit. While the computer game might demand your attention and rely on how quickly you respond, the personal tracking device just sits there, monitoring and counting the steps you take, for you to access and reflect upon at a later stage. The computer game might demand the use of a dedicated gamepad for interacting, while the personal tracking device only relies on what its accelerometer registers, no matter what shoes you wear. These two examples clearly illustrate very different modes of interaction and very different materialities for the interaction.

As we surround ourselves with more and more digital services, artifacts, and gadgets, these services and artifacts will also stand in relation to us as users in many different ways—and be manifested in many different material forms.

When we think about this growing landscape of interactive services and devices and to what extent we can now open up interaction design across a multitude of materials, we can consider design across these substrates from an *interaction dynamics* point of view. What do I mean by "interaction dynamics"? Well, dynamics is a key aspect of any interaction design.[12] Through our interaction with an interactive artifact, the state of the artifact changes. It is dynamic, and its dynamics are typically also reflected in changes in the user interfaces. So, when we design interaction using both digital materials (like code, scripts, and user interfaces in graphical forms on screens) and traditional materials, we also need to think about how we might use these traditionally noncomputational materials to be activated as part of the design, and how such materials can communicate this dynamics to the user. When using wood as a material, for instance, we notice that it can be bent, and we can think about the extent to which it can be bent as a design resource and as something that can be used to illustrate a particular state—a change or some dynamics—in the interaction design. Accordingly, wood can express a particular aspect of the "dynamics" in a tangible interaction design project.[13]

Wood is certainly not the only material that can be activated or reimagined as a computational resource. On the contrary, we can take almost any material and think about its properties from an interaction design perspective. When doing so, we engage in an act of reimagining traditional materials as computational resources.

For instance, if we take wood as a material, it has a number of unique properties that could probably be used as a computational resource. If we cut a few pieces of wood, we can notice how the patterns of rings are unique in each piece. This quality of wood can probably be used in computing: it might work as a natural tag (almost like a barcode or QR code), or a camera could be used to read the pattern as an identifier as well as to detect movement, etc. Furthermore, we can consider the moisture level of the wood (how dry it is, as a percentage) and we can probably think about ways of using that moisture as part of an interaction design project.

Iron is another example of an interesting material in this context. Iron is a strong material, but it can also lead electricity and heat. Concrete, on the other hand, is also a strong material, but it has isolation as one interesting property (thus it might be an interesting material in combination with iron). Another interesting material is glass, which is typically considered as a transparent material, but can also come in many different forms (blurred, frosty) and colors. Glass is typically considered a fragile material, but there are also special types of glass available that are strong (even bulletproof).

So far I have only discussed solid materials, but we should not forget about fluids and the possibilities such materials open up for interaction design. Just consider water as one such fluid material. Water can present itself in very many different forms, and in some cultures whole vocabularies have been developed to enable detailed descriptions of water in frozen form, including different forms of ice, snow, tracks in snow, and conditions of the use of snow.[14] In addition, water can appear both in fluid form and as a solid (ice); it can be free-flowing but also easily appears in the form of water drops. And this ability to form water drops can easily be eliminated if water is combined with other fluids, like alcohol, that destroy the surface tension of the water.

Icehotel, located in the north of Sweden, close to the Arctic Circle, is a hotel completely made of various forms of frozen water (including both snow and ice) as its basic material. In fact, Icehotel has made it their trademark to focus on only one material for their design and architecture; in addition to the basic structure of the Icehotel, just about all items inside the hotel (including ice sculptures, tables, chairs, and even entire chandeliers) are completely made of ice.

While the Icehotel is made out of a single material[15] and as such serves as a demonstration of a material-centered approach in the area of architecture, there are also important transdisciplinary lessons that can be learned from the way its designers approach this single material and how they do material explorations and design projects. On a general level, the main takeaway for interaction design is that the more we focus on and explore a material and experiment with it as the central element of a particular design, the more things we realize we can do with that particular material. In short, a narrow focus on a single design material will not lead to a narrow design space. On the contrary, a narrow focus might open up the design space for new ways of doing things and new design opportunities. The Icehotel case serves as a thought piece that allows us to think about what a material-centered approach can offer in terms of guidance for design, and an approach for exploring the potential of a selected design material.

If we now turn to interaction design, we can similarly consider how we can explore different aspects of interactivity by making use of different material properties in our interaction design projects.

Let us take as an example the growing interest in so-called water displays. By focusing on the properties of water (how water drops can be formed, how water drops can be separated from each other, and how fast water drops fall if released from a certain height) and thinking about these properties in relation to gravity and computation, we can see the main idea behind water displays (see figure 5.4).

As further described by Wiberg and Robles (2010):

Consider water displays. Taking advantage of already present structuring features of surface tension and gravity, these systems render images in falling water drops.

Even infosthetic versions like the Bit.fall installation, in which synchronized magnetic valves control water drops, allowing users to translate digital images from the web into tangible, if momentary images, are not technologically impressive inventions. Yet, it is likely that water could not be conceived through the features of discrete drops in flows unless viewed in a computational moment. (Wiberg and Robles 2010)

Figure 5.4
A computer-controlled water display similar to the *bit.fall* installation designed by
Julius Popp.

What the *bit.fall* water display project illustrates is basically interaction
design across different materials and substrates.[16] In *bit.fall*, computing is
combined with water and gravity as yet one more material. By understanding how gravity works and how water can be controlled to fall in the form
of water drops when released, it is possible to design a "display" that literally flows in mid-air.

On a more general level, this project also illustrates that a material-
centered agenda broadens the spectrum for how interaction can be manifested in the world (beyond traditional input/output modalities including
keyboards and screens). Far beyond any traditional interaction modalities,
the material-centered approach to interaction design sets no restrictions for
where the computer "starts" or "stops," the computer is no longer a gray
"box" on a desk, and it is no longer helpful to only think about computing
from the perspective of computer hardware and software. Far beyond any
such limiting perspective on computing, the material-centered perspective
acknowledges that computing can take any material form, that computing
can involve any materials, and that interaction design becomes a matter of
putting together different materials in meaningful ways.

To take just one more example of the current movement away from
"the box" and toward this form of interaction design, we can consider the
Runkeeper smartphone app for a moment. Runkeeper monitors the distance the user has run, his or her pace, etc., and plots the run on a map.
Designed from the viewpoint of traditional input/output modalities, Runkeeper would have been an app where the user had to enter data about the

run taken, the distance, the route, and so on. Probably this data would have been entered via keyboard and mouse if we had stuck to the traditional input/output modalities. However, we notice right away how clumsy this solution would have been, and we also know that this is not how interaction with Runkeeper happens. Instead, Runkeeper relies on the activity being done (that is, "running") and that activity in itself is also the main input modality to the app. The app works in the background; while the user runs, it automatically collects data about the pace, distance, and route for that particular run. Accordingly, the pace of the run is also data for the app to process, and so is the physical location of the runner throughout the run; the pace and location of the user of Runkeeper are also part of the interaction with Runkeeper, and part of the computing act that can inform the user after the run ends about how the run went. In short, the run is also part of the computing, and is the main interaction modality.

Clearly, different materials or parts, and the combination of these different parts, enable different opportunities for interaction design. However, to design useful interaction is not only about "uniting" different materials; it is also about balancing the inter-actions. Dialogue-based interfaces assume a turn-taking mode of interaction—that is, a well-balanced mode of interacting back and forth between the user and the computer. Scripts are a different design element that enables a form of interaction that tilts toward letting the computer do more, as one entity of the interaction, while the user can wait for the result/outcome of that scripted activity. In the case of Runkeeper, the direct mode of interaction is the pace and geographical route of the run, whereas the scripts running in the background help the app to monitor the run. Here some aspects of the interaction design are about active interaction while other aspects run simultaneously in the background. This configuration is enabled by combining these modes of interaction with GPS technology that can track the user's geographical movements with a mobile device and with a script that constantly collects data about the user's movements during a run. Overall, and from an interaction design perspective, this is about bringing the different pieces together, and accordingly it is about designing and defining the part-whole relations of the design.

Part-whole relations in material interaction design

One approach to the design of interaction is to think about *the form of the interaction*, as introduced in the beginning of this chapter. Thinking about form means considering how the user will interact with the interactive system or artifact and how the user will behave during the interaction. It is also about how he or she will experience a particular form of interaction, and how others will perceive this person during an interactive session. Another option is to think about the design of interaction from the viewpoint of *the session being designed*. The session might be designed to be short, involving only a few interactions, or to be extensive, both temporally and in terms of how the session is kept up and running across many different applications, devices, and other computational resources.

In both the first and the second approach, the design might be focused on different materials and how those materials might be used and integrated in order to manifest a particular design. This "particular design," or "unit of design," refers to something that is integrated and that we can put a label on—in most cases, in the context of material-centered interaction design, this will mean an object. Furthermore, a unit is something that works as a container for something and that holds different parts together—in short, a whole. If we can form an understanding of what an "interactive whole" is, what are the implications of this concept for interaction design? In this section I will address this question.

In the previous chapter, I introduced the notion of "relational practice" as a term for talking about how the practice of interaction design is increasingly about crafting and manifesting relations. The crafting of these relations stretches from the integration of different materials into computational compositions, to the crafting of how such computational compositions should or could relate to their user. In the field of human-computer interaction, we have developed a whole range of approaches to examine and craft good and meaningful relations between users and machines—including other relational approaches to interaction design, such as usability, ergonomics, contextual inquiry, and user experience design.[17]

Not only is the practice of interaction design "relational," but so is the unit of design.[18] Thus, an interactive "whole" can be defined as a system of interlinked parts that enable a user to interact with it through its predefined interaction modalities. This notion of "relational" is, however, not just restricted to the inner logic, or internal configuration, of how the different materials, elements, and subsystems are interconnected to support a particular form of interaction; "relational" is also a label for how the interactive artifact stands in relation to its user and to the interaction that unfolds during use. Through interaction, the state of the interactive artifact changes. This is part of what is meant to say that something is interactive—it responds to actions and it is, accordingly, part of *inter-actions*; that is, it is part of the actions taken between a person and an interactive object. As mentioned in chapter 3 and earlier in this chapter, this quality has also previously been referred to as the "dynamic gestalt" of an interactive system or artifact.[19]

As we set out to explore the relational character of interactive artifacts and how interaction is enabled through the design of thoughtful and meaningful part-whole relations, it becomes obvious that we also need a well-developed language that can enable precise discussions of part-whole relations in interaction design. We need a vocabulary that speaks to the "relational aspects of design." While much work still needs to be done to develop and establish such a design vocabulary, there are already a few terms in place that speak to some relational aspects of interactive systems. These terms include the notions of "integration," "interaction," "transaction," "processing," and "input/output." Each of these terms speaks to a different relational aspect of just about any interaction design. With those concepts as a start, we should be able to further extend and develop this vocabulary.

Developing this vocabulary will be challenging. One of the largest challenges is that our practice is increasingly focused on the design of even bigger, more complex, and more integrated systems, platforms, and digital ecosystems. However, while this ongoing development might cause difficulties, it only makes the development of such a vocabulary more relevant. As the relational aspects of interaction design scale from the smallest and simplest interactive device to the largest, most complex and integrated systems that enable interaction across a multitude of platforms, devices, and services, we need a vocabulary that allows for detailed discussions about design across this full range of systems. As I will discuss in the next chapter, it is partly this increase in complexity that calls for compositional approaches to interaction design. As formulated by Baldwin and Clark (2000, 63): "the essence of a complex thing is that its parts are interrelated. The different pieces must work together, and the whole must accomplish more than any subset of its parts." The important thing here is that the different pieces and parts are so aligned, so united, so intertwined, and so entangled that they come together as a whole, present themselves as a whole, and thus manifest themselves not as parts, nor as parts with connections between them, but as *one* thing, *one* design, *one* interactive whole.

User interfaces, whether they are graphical, tangible, or "faceless," need to be carefully designed, because it is through the design of those interfaces that users experience the digital artifact either as an integrated whole, or as more or less loosely coupled parts and pieces.

When working with part-whole relations in interaction design projects or interaction design across different materials, two possible strategies can be applied. Generally, an interaction designer can either do *within-material design* or *across-material design*. Both approaches are important for any material-centered project, but they serve different purposes in the design process. "Within-material design" refers to the in-depth exploration of a particular material and how it might be used in the design, whereas "across-material design" refers to the activities and inquiries that need to be done on how a particular material might work in relation to another material, and how one material might affect or make use of some properties of another material.

To fully address the relational aspects of doing material-centered interaction design, one must explicitly focus on the materiality of interaction, including both the materials and the relations between different materials, in order to explore new grounds for interaction. As stated in Wiberg and Robles (2010):

Reconciling the division between physical and digital means not only crafting metaphorical relations, like GUIs, nor even enabling physical analogues for digital information, like TUIs. Complementing these approaches must be a design space for broadly imagining what kinds of new materials and relations between materials are possible at a range of depths—from interface to structure—and at a variety of scales—from objects to architectures.

It is not only at the level of large scale-structures—whether architecture or large-scale digital platforms—that we face challenges for how to craft good part-whole relations in the area of interaction design. On the contrary, even when designing the smallest interactive object, it is important that the materials and digital services are integrated in a way that is not only tangible but also meaningful. The next chapter will explore the current development of "smart rings," and more generally the trend toward "digital jewelry," as an example where compositional thinking is crucial for arriving at meaningful and hopefully beautiful designs.

While the goal is clear—to design good compositions—it is not equally clear how to do so. We need approaches, methodologies, and not least experienced designers that can support and guide our thinking on compositional interaction design.

This chapter started out with a discussion of material properties and properties of interactivity. This discussion formed the groundwork for thinking about and discussing part-whole relations in the context of interaction design, and enabled us to arrive at a state where we can start thinking about compositional design in the context of interaction design. But what would a compositional approach to interaction design mean in terms of design thinking? And can we sketch out a design philosophy for interaction design that rests on a compositional ground as an approach to design? And finally, what might the methodological implications be if we apply a compositional approach to interaction design? These important questions will be addressed in the next chapter.

6 Interactive Compositions
Compositional Thinking Meets Material-Centered Interaction Design

Interaction design is a relational practice[1] in which interaction designers bring many different materials together into compositions that enable particular forms of interaction. However, even though interaction design is about the design of interaction, interaction as an act that unfolds between a user and a digital artifact cannot be completely "predesigned"—that is, it cannot be fully defined and controlled from a design point of view. Users might use a piece of technology as intended, but sometimes technologies are also used in unintended ways. Therefore, the interaction designer can design not the interaction, but good preconditions for a particular form of interaction, a particular materiality. When this materiality is used during acts of interaction, we can refer to this intertwined relation of interactions enabled by, and performed with and through, material configurations as "the materiality of interaction."

The materiality of interaction is thus the material footprint of compositionally arranged materials that allows for a particular form of interaction to be performed. As I highlighted in the previous chapter, this design work of forming material compositions with a particular type of interaction in mind can be further informed by thinking compositionally along a number of interaction design dimensions. In the previous chapter I introduced six such interaction design dimensions. To recap, we can think about the design of interaction as a matter of defining (1) the *changes of states* (*dynamics*) in the interaction being designed, (2) the speed of change (*pace*) of the interaction, (3) requests for input (or the level/balance of turn-taking in the interaction being designed), (4) the *responsiveness to input* (*receptiveness*), or the extent to which the intended user can intervene in the flow of interaction, (5) whether the interaction is *single-threaded or multithreaded*, and finally, (6) to what extent the interaction is a *direct* or *an agent-based* (*"invisible"*) *interaction*.

Clearly there are many aspects and dimensions to take into account when designing interaction (in addition to thinking about how to materialize its intended form). In this chapter, I present a set of ideas for how an interaction designer can think about, and do, compositional and material-centered interaction design.

Philosophy—On compositional material design thinking

First of all, and in order to talk about *compositional interaction design*, or *compositional material design thinking*, we need a basic understanding of what we mean when we talk about *composition*. How do we define this notion of "composition" in interaction design? After elaborating on this definition in this section, I will move on to how a compositional approach to interaction design can be not just a method or procedure, but also can work more fundamentally as a conceptual ground for thinking about the design and materiality of interaction. In short, this chapter is about the philosophy behind a design-oriented, material-centered, and compositional approach to the materiality of interaction.

But first, let us see how "composition" is understood and defined in some related areas of design, such as the fine arts, architecture, and music. Within the fine arts, "composition" refers to the specific way the elements of a work of art have been arranged in relation to each other. As such, composition in the fine arts is a relational notion that describes the order of things, where "order" can be thought of as the defined interrelations between things. In music, on the other hand, "composition" can refer to the work itself (the musical composition), but it can also be used as a verb—*to compose* music. Over the years, many attempts have been made to theorize composition as a relational concept that speaks to the arrangement of parts into wholes, with a particular focus on how things are interrelated. "Composition" as a noun has also been theorized as the label for activities aimed at combining, blending, and bringing together pieces into larger structures—that is, into compositions.

In an attempt to theorize composition, Goldstein (1989) developed a vocabulary that he refers to as a set of the principles of composition (including the notions of balance, emphasis, simplicity, hierarchy, and unit). His proposed vocabulary was meant to provide a way of talking specifically about various aspects of the visual arrangements of things. To add to this vocabulary, Ian Roberts (2007, 13) focused on "armature" as "the fundamental lines of direction or flow that connect the main compositional movement to the picture plane." Roberts thus added movement as a dimension of composition. He was also interested in how our perception of things frames our understanding of a composition, commenting: "Seeing composition in terms of shapes and flows is not an intellectual idea you apply; it's a perceptual shift" (Roberts 2007, 8). With this idea as a point of departure, he also focused on cropping and framing—two techniques that also speak to the relation between the composition and the observer of the composition. Henry Rankin Poore (1976) has pinpointed that it is not only the spatial arrangement of things that influences our understanding of (visual) compositions, but also other elements such as light and shadow.

Composition is clearly about the formal arrangement of different elements into functional and aesthetic wholes[2]—and to a large extent, the literature on composition mainly focuses on visual compositions within the arts, in photography, and in architecture. But from the viewpoint of interaction design, there are additional elements at play, ranging from the (almost obvious) visual elements, to temporal elements, to functional elements, and to the different materials that manifest the design.[3]

Let us start with the visual elements. In many interaction design projects, the interactive surface is screen-based. Here the arrangement of visual elements is crucial. This is obvious when considering the design of a graphical user interface (GUI), but it is also true for other ways of making the interface visible. Questions of appearance and the importance of deliberately defining what a particular form should communicate via its visual markers have always been important to artists, product designers, industrial designers, and architects. In many cases, the visual presentation sets the frame for how a certain product, building, or piece of art should be interpreted and made sense of. Sometimes storyboards and wireframes are used as tools to help the designer decide in which order different visual elements should be presented, and what the relation should be between different visual elements.

The order in which the visual elements of an interactive artifact are presented is first of all a matter of *interaction logistics*. As an example of interaction logistics, in just about any traditional online store the user needs to follow certain schemas or logical steps of interaction in order to use the site. One first needs to find and select a product, then click "Add to cart," then click on "Cart," then click on "Proceed to checkout," and finally go to "Pay." The interaction order is also closely related to the *temporal* elements of any interaction design. Whether the temporality is related to the adjustment of speed—as in many computer games, where changing the speed makes each level slightly more difficult—or it is related to some information processing the computer needs to do or data that needs to be stored or retrieved from a database (while keeping the user waiting), temporality is something that can be designed. These temporal elements (interaction pace, speed, timing, waiting time, etc.) then become part of the composition being designed.

But interaction design is not only about visual presentation or the pace of the interaction; it is also always about what the interface offers in terms of functionality. In fact, I cannot think of any interactive system that fails to offer functionality. Accordingly, the design of the functionality is also an essential element for any interaction design project. Here, the functional elements of an interaction design work as "the glue" that holds the other elements together.

Let me introduce a simple example to illustrate this. A link enables the switch from one webpage to the next, and an "undo" button enables a user to go back to a previous state in the interaction. Furthermore, functional elements are the components of an interaction design that is related to the output of the interaction. Whether the output is instant (for instance, the computer instantly writes an "A" on the screen if the key "A" is pressed on the keyboard) or it comes at the end of the session (for instance, the product order is sent, or the document is printed), the arrangement of the functionality in relation to the visual and temporal elements of the interaction is crucial for a well-designed interactive composition. Again, this functionality is not only a matter of what is technically implemented; far beyond that, it illustrates how interaction is highly intertwined with how it is manifested. In short, the materiality of interaction depends on the details of how it is implemented. As formulated by the famous furniture designer Charles Eames: "The details are not the details. They make the design."

Returning to the area of interaction design, Steve Jobs once said that "interaction design is not only about what it looks like, but more fundamentally, it is about how it works."[4] With these words he highlighted that interaction design is always about understanding the visual in relation to the functional, and that even the visual needs to be understood from the viewpoint of the functional, that is, "how it works." Jobs pointed toward how a design worked during use—that is, how it worked in relation to an intended mode of interaction—and addressed how well the material composition supported a particular form of interaction.

In this book I propose that we can open up such detailed conversations about interaction design compositions so that we can discuss not only functionality or technology, but more fundamentally preferred forms of interaction, suitable material manifestations of interaction, and how a particular form of interaction might be manifested through its materiality.

In fact, these three aspects of interaction design compositions—the visual, the temporal, and the functional elements of interaction design compositions, as well as how these elements are interrelated—can serve as a cornerstone framework to analyze existing interaction designs. We can use them to evaluate specific designs, but also for the more long-term purpose of building a rich repertoire of examples of different types of interactive compositions. Such rich repertoires can be important in order to have a set of models of interaction design compositions in mind when starting up a new interaction design project, or when exploring different alternatives during an ongoing project.

In relation to the overarching theme of this book—the materiality of interaction—I see these three elements—the *visual*, the *temporal*, and the *functional* elements of interaction design—as important cornerstones for thinking about how interaction can be manifested in material form. Even though the graphical theme is set, and the order, pace, and functionality are defined, there are still many decisions that need to be made in relation to how a particular form of interaction can be manifested in material form. For instance, if the interaction is mainly screen-based, decisions can still be made about screen resolution and screen real estate, whether the interaction with the screen should be keyboard- or mouse-based or via touchscreen, and so on. These issues are all about material manifestations, the intended form of interaction, and accordingly, the materiality of interaction.

Even if the functionality is decided upon, the materiality can be very different. Each aspect of how the functionality is implemented and materially manifested in the design will also have a direct impact on the interaction model and on the user experience while interacting with the system or artifact. As such, the material aspects of the interaction design not only provide a container for the other elements, but also relate to how the interactive system will be perceived, culturally interpreted, and valued. More importantly, the materiality affects the user experience along the full range of aspects from "look and feel" to the symbolic value of the artifact—both from the user's viewpoint and based on the way others perceive and interpret the design as part of a cultural and social context.

So what are the implications of this line of reasoning for a foundational philosophy for doing material-centered interaction design? First of all, it becomes obvious that creating good compositions across these different elements cannot happen without a clear idea about what kind of interaction one wants to design. That is, the intended form of the interaction being designed needs to be carefully considered in order to be carefully crafted. In order to combine visual, temporal, and functional elements into a composition that constitutes a well-working interaction design, one must have first thought about the form and materiality of the interaction being designed, and then used that as the guiding idea for bringing different materials and elements together into a formal composition. One cannot start with only the visual elements or how the artifact should present itself, and then try to "add" the other elements later. When composing interaction, the interaction designer needs to move dialectically between these elements and back and forth between the materials, the parts, and the whole in order to bring the composition together. For this reason, the design process is not linear but highly dynamic, and it involves many parallel activities, such as comparing different design alternatives; evaluating particular materials, elements, and parts; experimenting with alternative materials and elements; and even designing new parts.[5]

Composing interaction is an act that reaches beyond the creation of a single element. Beyond being just the development of a function or a visual element (although these can certainly be part of an interaction design), the compositional act of interaction design means integrating and bringing together existing elements and adjusting these elements to work together as part of a whole.[6] Furthermore, this "whole" will in many cases be part of an even larger whole, as most interaction designs nowadays are part of larger systems, platforms, or other unifying structures that both enable and restrict the particular composition being designed. Rather than getting into a discussion of the design of systems, subsystems, and even infrastructures, however, this book will stay focused on interaction design and on the inter-action being designed. In this context, "compositional interaction design" refers to the act of designing interaction by bringing together different materials into a composition that will enable that particular interaction; in addition, the imagined form of interaction must work across the whole composition being designed.[7]

This compositional act of integrating different parts and elements into a whole requires a good understanding of the whole that is being designed, which in turn demands a good understanding of the interaction being designed and how that relates to its material instantiation. Not only does the interaction designer need to have a strong understanding of the whole, of the interaction being designed, and of how the interaction can be instantiated, but he or she also must have a strong knowledge of the materials, parts, and elements that will be part of the overall composition. Selecting these different materials, parts, and elements demands knowledge about existing possible choices as well as the skills needed to develop new parts that don't yet exist, the ability to evaluate and make judgments about different materials, parts, and elements, and knowledge about how the ones that are selected can be brought together in a functional interactive compo-sition. Furthermore, the creation of new parts also demands imagination, creativity, problem-solving skills, and an explorative mode of inquiry.

As I have illustrated here, compositional interaction design assumes and requires a number of quite different design skills. Beyond the skillsets needed to work with the single elements and materials in interaction design—including the wide range of skills from graphical design, to programming, to hardware development and configurations, to knowledge about different materials, and so on—the compositional approach to interaction design also requires a number of additional and quite different skillsets.

To define what those additional skillsets might include, we must keep in mind that compositional theory teaches us that a composition relies not only on the *part-whole relations* that constitute the composition (which demand both skills related to the parts, or "modules" of the design, and skills related to the crafting of the whole, which is here thought of in terms of its overall *thematic design*), but a well-working composition also needs to rest on a *clear philosophy* about how the different parts that constitute the whole are interrelated or interconnected. In the context of the materiality of interaction, I think about this in terms of the composition's *unifying design.* The whole composition also needs to be *anchored* in different ways—in relation to other design ideas, and in relation to similar or different ways of composing. In short, the interaction designer needs skillsets related to motivating and grounding the design in terms of its *underlying philosophy.*

First of all, the skillsets needed for working with "thematic design" in the context of compositional interaction design thinking refers to skills necessary for composing, compiling, and presenting an overall theme for the interaction design. This theme is not to be thought of in terms of a graphical schema (although such a schema could be aligned with the thematic design if part of the interaction is graphical/visual). More fundamentally, the thematic design has to do with the form and materiality that the interactive product or service should manifest and express. By "express," I mean that the final product can also work along a full scale of different design opportunities.

Some thematic designs rely heavily on visual presentation and very visible modes of interaction; the Microsoft Kinect platform, which includes both screen-based interaction and gestures as the main interaction modality, might work as an illustrative example. On the other hand, an example of less visible, or even "invisible," modes of interaction might be the smart tag on the dashboard of your car that communicates with the road payment system to pay the toll for using certain roads or bridges. In short, a theme conceptually brings together into a meaningful whole the different concepts on which a particular design relies. A *theme* is the central idea behind a design, and the structure that enables its user to understand single operations, at any stage of the interaction, in relation to how the interactive artifact, system, or service is thematically structured. To organize everything into an overarching theme demands skills related to the abilities of abstracting, generalizing, and concluding. In short, the act of doing "thematic design" demands skills related to making abstractions and making sense.

While the skillset related to doing thematic design involves the ability to make abstractions, the designer also needs the skills to organize loosely coupled parts into modules or design elements, while at the same time keeping a keen eye on how the different modules will work as part of the overall composition, and how each module will reflect the thematic design. The skillset needed for doing such "modular design" work thus includes the abilities and skills for working back and forth between the details and the whole, and for working back and forth between the modules and the materials so as to iteratively make sure that the thematic design is reflected all the way back to the selection of particular parts and materials for the design. Skills are also needed the other way around, to make sure that the overall design stays true to the meaning and symbolic values communicated by the single materials selected and included in the design. This ability is crucial for the design to appear authentic rather than kitschy; a typical design mistake is that the designer tweaks a material into something it is not, or compromises the design with materials that communicate something it is not. Compositional material interaction design is also very much about presentation. It is not enough to merely bring the different pieces together; the pieces must be presented as a whole. This is why design skills for doing thematic design are so essential.

Finally, while the skillsets related to "modular design" are also about reducing the complexity of larger compositions into manageable parts or subsystems, designers also need skills related to knowing how to bring the different modules, parts, and elements together. In the context of interaction design, this skillset might include knowledge about particular application program interfaces (APIs), ways of calling different functions or procedures, ways of "hacking" otherwise inaccessible data, or ways of using various sensors and actuators to work across digital, analog, and smart/dynamic materials. These skills are practical ones, including the actual programming to get the pieces to work in concert, but it is equally important to make sure that the integration is done in a structured way, and preferably done in the same way across the whole implementation. In order to do that, it is crucial to know how such structuring is best done for compositional interaction design.

All of these different skillsets are related to the practice of doing compositional interaction design. The next section of this chapter will focus in more detail on what compositional and material-centered interaction design might imply in terms of overall methodology.

Methodology—On doing compositional material interaction design

In order to do compositional material interaction design, we not only need to know what it is (which is one of the reasons that we need a philosophy behind this approach, including clear definitions and a theoretical ground), but we also need ideas about how to conduct it in practice. In other words, we need a good understanding of the process of doing compositional material interaction design, and given such an understanding, we might be able to formulate some aspects of that process into a methodology and method. This section is a first attempt to develop a method and methodology for doing compositional material interaction design.

In sketching out a methodology, we can start by identifying some different aspects of this area that are essential. One of the most central might be this focus on materials, but also on material properties and how materials can be brought together and combined in different ways. There are a number of additional things that are important when doing compositional interaction design—not least, "what we do" with these materials and the different activities we undertake when designing compositional material interactions. In this section I will talk specifically about activities related to exploring, experimenting, selecting, integrating, composing, interacting, discussing, and reflecting on or with these materials.

The activity of *exploring* and *experimenting* with materials includes not only getting to know the materials' physical properties (although that is important) but also material programming[8] and experimenting with how the activation of one physical material or substance can be used in the further activation of other materials (physical, digital, or even immaterial substrates).

For instance, different materials have different properties that can be useful in a design project. Just to mention a few examples, we can think about how wood can be bent, plastics can melt, and iron can not only be heated up, but can keep heat quite well over a period of time. This last example illustrates that temporality is an important factor to consider in relation to material properties. For instance, although iron can hold its temperature over a certain amount of time, it also takes time for iron to cool off, and it takes time for iron to heat up. Accordingly, material resistance is a dimension that needs to be considered when doing material-centered interaction design. Materials can also be conductive or nonconductive (a quality that can be valuable for an interaction design project), and different materials might even come with some basic functionality (for instance, smart materials like "memory-shape alloys" can "remember" a certain form and can thus allow for shape-shifting designs). In short, understanding a material from the viewpoint of material-centered interaction design is by no means only related to the understanding of material properties. Far beyond that, it is about understanding how a combination of different materials can work as the instantiation of the materiality of the interaction being designed.

After materials have been explored and experimented with, we can move on to *material selection*. This activity involves carefully evaluating a wide range of different materials and different material constellations that might solve the current design challenge, although in slightly different ways, and deciding which materials to include in the final design and which to exclude.

Having explored and selected the set of materials to be used in the design, the activities of *material integration, composing*, and *craft* take place. These activities include the practical work of programming the materials and bringing software, hardware, and solid or fluid materials together. At this stage, the pieces are put together into a functional and compositional whole and made to appear as a whole. Consequently, these activities require working back and forth between the materials, the parts, and the overall design—the whole—of the interactive system.

Having reached the stage where the materials are integrated and the system is "up and running," there is still an important step left—engaging in the activities of *discussion* and *reflection*. Discussions allow for a social exploration of the final design. This is a critical activity aimed at examining the design from a multitude of viewpoints, and evaluating whether the final design met the design goals and whether there are opportunities for further improvements. Reflection is also a critical activity, aimed at thinking through the final design not only in terms of how it is configured, but also whether it addresses the initial, identified design challenge. Perhaps even more importantly, this stage focuses on thinking through the whole project in order to understand what can be learned from it, and what can be the takeaway for the next project (or for continued designing related to this particular user need, design problem, or use context).

In a recent paper (Wiberg 2014), I discuss a number of these activities under the umbrella of a "methodology for materiality." This is a methodology that highlights some important steps for doing compositional and material-centered interaction design.

In the next section I will present three different cases that serve the purpose of illustrating what compositional material-centered design looks like in practice. Rather than suggesting that material-centered design is something we as designers should do, I argue in this book that we already have this practice, and that we can become even better at doing it if we also articulate how it is done. This book serves that purpose of articulating and reflecting on this practice.

Turning to practice—Three cases and design projects

In the previous section I outlined the methodological ground for a material-centered approach to interaction design and I demonstrated how that could work as a frame for a set of design activities—including the *exploration, experimentation, selection, integration, composition, interaction, discussion,* and *reflection* on and with materials—in interaction design projects.

On a methodological level, this approach looks like a streamlined process, from exploring materials all the way to reflecting on the entire project, but what could this process look like in practice? To answer this question, I will in this section present three short examples and cases to add to our understanding of this matter.

Example 1: Composing with only one material

From 2006 to 2008, I led a research project in collaboration with Icehotel in Jukkasjärvi, located in the far north region of Sweden. Icehotel is, as its name implies, an entire hotel made solely out of one material—frozen water. Every winter, the Icehotel is built with a new design, and every spring it melts—the frozen water turns into water and flows back into Torne River. Icehotel has been around for more than twenty years now, and each year its designers and architects have worked intensively to explore new forms for the hotel—both for its interior and exterior design (see figure 6.1).

Icehotel is a living example of what material explorations looks like in practice. By staying close to one material and viewing its material properties not as a design restriction but rather as a design opportunity, Icehotel has demonstrated how a material-centered approach to architecture and interior design can work as a stable ground for creative material-centered design.

Icehotel is, as such, a good example of how one can move from a particular material to a complete design, and it demonstrates the potential of a material-centered design approach.[9]

a.

b.

Figure 6.1
Two illustrations based on pictures from the Icehotel that show the exterior (top) and the interior (bottom) of the hotel. Everything is made of the one material—from the floor to the ceiling, from the chairs to the chandeliers.

Example 2: Composing across physical and digital materials

As described above, one challenge facing the material-centered design approach is how to work compositionally across different materials and to think relationally across the different parts of the design. This challenge is faced by any area of design, not just the field of interaction design. While this general difficulty could be seen as something problematic, it also opens up opportunities to learn from other areas of design, and to gain inspiration for how to move forward with specific design challenges. For instance, the field of fashion might help us to understand some aspects of how to design digital jewelry.[10] To illustrate this and to exemplify how interaction design nowadays stretches across physical and digital materials, I will take the contemporary growing interest in a particular kind of digital jewelry—"smart rings"—as an example here.

"Smart rings" are rings that people can wear on their fingers—just like ordinary rings—but with built-in computational capabilities. For instance, a smart ring can vibrate if you have an incoming call or a text message, and some current smart rings can also lock or unlock your phone, share and transfer information, and control some apps. In terms of its visual appearance and material instantiation, it is a ring—literally speaking (see figure 6.2).

In terms of its form factor, this computational object moves beyond any skeuomorphic interaction design. The smart ring is not just "metaphorically inspired" by the physical world (in this case, the ring as the physical object in focus); far beyond that, the smart ring borrows the actual form and symbolic meaning of a ring as a platform and point of departure for its materiality of interaction. The decision to give this interactive artifact a physical form also places it in the world, and on the finger of its user. As such, it is a radical interaction design with a materiality that not only supports some particular forms of interaction, but also occupies physical space.

However, while it is clear that this interactive artifact relies on the ring as its overarching form factor, it is less clear how the interaction with this device is configured. This digital artifact has a number of digital services attached to it, but how are these services made available to its user? An observer might wonder whether the different services are available by simply turning or stroking the ring, or whether there are other, even more sophisticated ways to accomplish "human-ring interaction."

Figure 6.2
The ChiTronic Smart Ring is an example of interaction design across physical and digital materials.

In terms of its form factor, the material form of the ring works—this "smart," interactive device clearly looks like a ring and is interpreted as a ring. But the embedding of computing in this particular material form, this literal way of defining the materiality of interaction as a "smart ring," also comes with a set of assumptions about how it might be interacted with. Accordingly, when interaction is designed across physical and digital materials, the resulting materiality not only serves the purpose of manifesting the design in a material form, but that form also communicates, symbolizes, and in other ways is associated with particular ways of interpreting the design, which in turn will define how it will be used and interacted with. The materials give a particular interaction design its materiality, and in turn the same materiality also to a large extent defines the interaction.

Example 3: Designing modules and composing wholes
My third example of what a material-centered approach can look like in practice is related to the compositional design of large and complex systems that might include many parts and subsystems.

When complexity increases, part-whole design becomes more than just a matter of combining materials into larger wholes. While that is still necessary, a middle step needs to be introduced to the design of different parts, elements, or subsystems. Here, these parts, elements and subsystems are referred to as "modules."

Compositional design with modules is about two things: the design of the different modules, and the design of the interfaces that enable the modules to form larger structures and systems—that is, to make sure that the modules work in a compositional whole. Here I use two pictures from a research project we ran in collaboration with Electrolux in 2010–2011 to illustrate the relation between different modules and the whole in which they play an important role (see figure 6.3).

Modular design thinking[11] is crucial for a compositional approach to interaction design. In many cases, interaction design projects are large, and an interactive artifact (say, a smartphone) can be a part (or module) of a much larger platform (for instance, the telephone network, the app store, and the ecosystem of devices in which the phone is just one module, and so on).

From the perspective of material-centered design, compositional thinking in the form of modules enables the integration and interoperability across different parts. We can think, for example, about the physical design of the "home button" on the iPhone that originally had the frame of an app as a symbol on the physical button. By pressing the button, the user was returned to the home screen on the iPhone, and as such the button worked as a physical element that took the user back to the home for all the apps on the phone. This demonstrates a link between one element (the physical button) and its particular place in the overall interaction model for the iPhone. From that location, the user could choose to go into any third-party app. (These different apps can be thought of as separate but still interlinked modules, with the home button holding the apps together and enabling the user to quickly go back to the home screen on the device.)

These three examples all illustrate the value of working compositionally in design, but what about interaction? What are the more fundamental implications for interaction design, given these three examples? Or to formulate this slightly differently: How can we "compositionally" design interaction? To answer these questions, we first need to go a little bit further into what we mean by this notion of "interaction" before we can move

a.

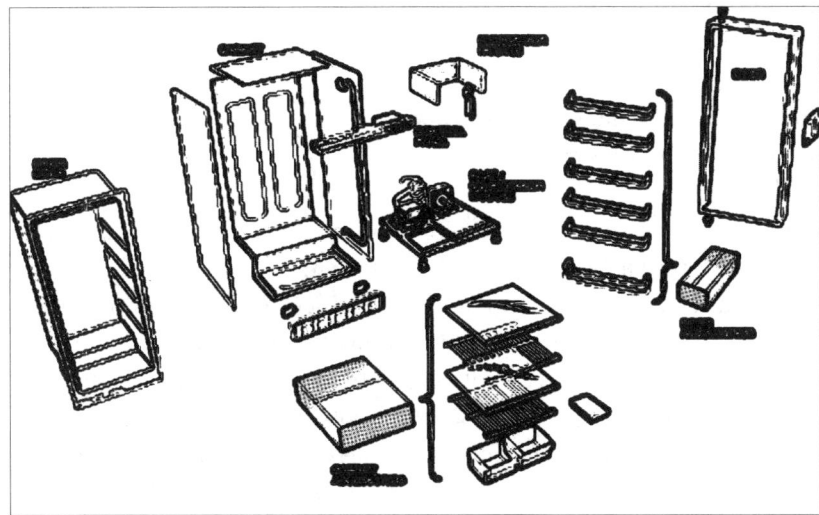

b.

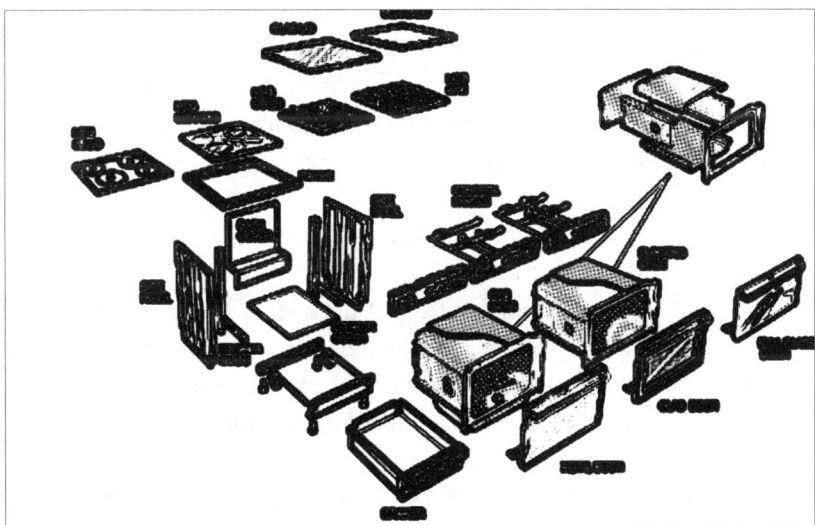

Figure 6.3
Two examples of modular design by Electrolux—a refrigerator (top) and a stove (bottom)—showing how each product is constructed from a set of basic modules.

forward and think about compositional design approaches to the materiality of interaction.

Toward a design theory of interaction—
Some basic questions and models

Before going into this topic, we need to ask a quite fundamental question:
Why do we need a design theory of interaction?

I could certainly go into a quite complex discussion here, but the simple
answer is that any theory is the articulation of how we see and understand
a particular thing, in this case interaction from an (interaction) design perspective. A theory helps us to describe an object of study, it gives guidance
in terms of what to focus on, and it articulates the implications of adopting
that specific perspective. For the area of interaction design, a good understanding of interaction is helpful in order to know what the object of design
is. In short, an articulated idea about interaction is helpful for doing good
interaction design.[12]

However, there are also a number of additional questions that need to
be raised in my striving for a better, or at least more theoretical, understanding of interaction. In this section, I will therefore shift between such
questions and then seek some answers through a review of the literature
in our field.

A good understanding of interaction—that is, having both some experience working with interactive systems and a well-elaborated theory of interaction—is not only good for intellectual/theoretical purposes, but is also
essential for the practice of interaction design.[13] However, for a design theory
of interaction, we first need to address how we are defining "interaction."[14]

I started to address this fundamental question at the beginning of chapter 3, but now I want to propose a theory and a model of the materiality of
interaction that will unify my understanding of interaction and "interactivity" with our field's increasing interest in "materiality." This fundamental
question about interaction deserves some additional reflection.

Dubberly et al. (2009) have raised three fundamental questions in an
attempt to start answering this question. These questions are:

- *"Do we agree on the meaning of the term 'interaction'?"*
- *"Has the subject been fully explored?"*
- *"Is the definition settled?"*

In their view, the answer to all three of these questions is "NO!" They then describe the current frameworks developed to address interaction, starting with the work by Davis (2008) and Buchanan (1998). The latter contrasts earlier design frames (a focus on form, and more recently on meaning and context) with another design frame, i.e., a focus on interaction. In doing so, Dubberly et al. (2009) build on Buchanan (1998) and suggest that interaction is the relationship between people and the objects designed for them, and that interaction is a key aspect of function. So here the suggestion is that interaction is a concept that is heavily focused on function and that it speaks to relational aspects between people and objects.

If we look at the wide range of literature in human-computer interaction (HCI) and interaction design produced over the last thirty years, we can see that a wide range of models has been developed to describe and analyze human-computer interaction. These range from models of the basic components of human-computer interaction (e.g., Benyon 2010) to more advanced frameworks and theories on the complex interplay between humans and technologies.

However, when we review this genre of HCI models of interaction, if only from a graphical perspective, it becomes obvious that most of the focus has been on humans and computers and very little attention has been paid to modeling the interaction per se (in these basic models of human-computer interaction, the entity of "interaction" is typically represented only as a thin line in between the human and the computer). This is also the case in the academic literature published in the area of HCI; most literature on HCI is concerned either with human factors or understanding human interaction with computers, or with the design of computer support to scaffold human activities.

The most basic models of HCI typically describe it as including the three entities of *humans*, *computers*, and the *interaction* between the two first entities. Such models are typically drawn as illustrated in figure 6.4, where interaction is typically reduced to some kind of arrow between a person and a computer, or to the form of a loop between the user and the computer via input/output flows (see, for instance, Heim 2007).

According to Dubberly et al. (2009), these canonical models of HCI follow a basic archetypal structure—the feedback loop. In their description of this basic structure, information flows from a system through a person and back to the system again. The driver of this loop, they argue, is a person

HUMAN   INTERACTION   COMPUTER

Figure 6.4
A basic model of human-computer interaction.

with a goal, and he or she acts to achieve this goal (provides input to the system), measures the effect of the actions taken (interprets output from the system), and compares the results with the goal. The outcome of this comparison then directs the next action taken by that person. Dubberly et al. point to Norman's "Gulf of execution—Gulf of evaluation" interaction model (Norman 1988), as well as his "Seven stages of action" model, and then outline three important additional questions in relation to these simple feedback-loop models of interaction:

• *"What is the nature of the dynamic system?"*
• *"What is the nature of the human?"*
• *"Do different types of dynamic systems enable different types of interaction?"*

While these questions might be focal, given the traditional interaction model paradigm that focuses on the human or on the dynamic technology, shifting the focus toward "the materiality of interaction"—that is, toward the enabling materiality of interaction, and how interaction also co-produces that materiality[15]—could lead to other models of interaction, and thus also raise other questions about interaction. For instance, we might pose questions more closely related to the character and dimensions of interaction and interactivity, as presented in chapter 5, rather than questions about system properties or human factors. As pinpointed by Dubberly et al., we still lack a good, shared understanding of what we mean when we talk about interaction.

As I stated above, there are, of course, a number of existing models of interaction already developed in our field. For instance, there are several examples of input-output models for HCI (to name just a few, see Marchionini and Sibert 1991; Boehner et al. 2005; Barnard et al. 2000), not to mention the human processor model (Card et al. 1983); models describing standards for designing good consistent interaction, including the WIMP-standard and other GUI standards (e.g., Dam 1997); and a number of models that describe different specific interaction modalities and interaction paradigms (e.g., Jacob 2006; Ishii and Ullmer 1997; Jacob et al. 1999; Abowd and Mynatt 2000; Bellotti et al. 2002). Furthermore, several recent models of users in relation to advances in the field of HCI include models of human perception of HCI (see, for instance, Kweon et al. 2008; Dalsgaard and Hansen 2008). Together with this work, there have also been attempts made to theorize interaction in relation to systems (see, for instance, Bernard et al. 2000).

In widening the scope of this literature review, I notice that there are a number of good models developed for informing design based on the analysis of basic and frequent human activities, including, e.g., the task-artifact model (Carroll and Rosson 1992), task analysis (e.g., Pinelle et al. 2003), and the reference task model (Whittaker et al. 2000). My literature review also acknowledged a number of workplace interaction models (for instance, Suchman 1995; Whittaker et al. 1997), computer-supported group collaboration models (e.g., Ellis et al. 1991; Whittaker 1996; Hindmarsh et al. 2000), coordination models (e.g., Dourish and Bellotti 1992), and models describing other social aspects related to HCI. Delving further into the area of social interaction in the context of HCI, we should also acknowledge models such as distributed cognition (e.g., Hollan et al. 2000) and situated action (Suchman 1987).

Moving from the social context of HCI to some area-specific models of interaction, we can find several successful attempts to model different aspects of interaction, including "embodied," "mobile," "tangible," and "social" interaction. For some additional descriptions of these models, see for instance the work by Klemmer et al. (2006) on embodied interaction, the work by Hinckley et al. (2000) on mobile interaction, the work by Hornecker and Buur (2006) on tangible interaction, or the work by Wiberg (2001) on social, "ongoing," and "seamless" interaction.

However, while these models help to categorize different kinds of interaction, they fail to address the basic question, "What is interaction?" In fact, adding another word before "interaction" in order to be more specific leads to a better understanding of that particular form of interaction while at the same time failing to describe, define, and model interaction per se.

From yet another perspective, the field of HCI has witnessed several valuable attempts to build theories of interaction through design. These include "proof of concept" approaches to interaction design (e.g., Toney et al. 2003) and the "artifact as theory nexus" approach formulated around the idea that artifacts can manifest interaction theories through their design (Carroll and Kellogg 1989) and that special characters of a certain interaction model can be explored through the design of a prototype (Lim et al. 2008). While such approaches might be useful for bridging theory and design when doing prototyping in HCI/interaction design, and while they might help to describe particular design concepts, they offer very little in terms of describing what interaction is in relation to the materiality that both enables and is formed by the interaction.

Clearly, a better understanding of "interaction" and a better understanding of the materiality of interaction could be a valuable addition to this current body of research on interaction, as well as for the profession of interaction design.

As we can notice from this literature review, several models have been developed that address and describe different ways of designing and evaluating computer support for interaction; several models have been developed for describing and analyzing human cognition, perception, needs, behaviors, motivations, activities, and goals for engaging in interaction with computers; and models have been developed to describe how humans might gain from interaction with computers in solving various tasks. So far, however, few attempts have been made to address the basic dimensions and aspects of interaction, and even fever to address how interaction and the materiality of interaction could be modeled. With this as a point of departure, in the next section I will sketch out a first draft of a model that focuses on the materiality of interaction in an attempt to describe how interaction works in relation to its users and its enabling (computational) materials.

Toward a theory and a model of the materiality of interaction

So far, this book has moved from an initial discussion of the representation-driven approach to HCI, via the material turn, to the introduction of a material-centered approach to interaction design. Any attempt to model interaction with this as a background needs to focus on how interaction can be modeled in relation to the materiality that both enables and forms that materiality. The model of the materiality of interaction constructed here addresses interaction from the viewpoint of a material-centered approach to interaction design. In chapter 5, I discussed how interaction design is a "relational practice," and as this model will illustrate, that description also holds true for the unit of analysis here—the materiality of interaction is fundamentally relational, and this relational character stretches all the way from the use of digital artifacts to the materials that make interaction possible.

I will return to an explanation of what I mean when I say that "the materiality of interaction is relational." However, I can say right away that this relational character has to do with how the materiality of interaction is formed by (or stands in relation to) both (1) *processes of use* and (2) *processes of computing*. In this model, I will refer to these two loops as "threads of interaction" (use) and "threads of computing" (processing).

Furthermore, it is my intention that this model of the materiality of interaction should acknowledge the need for *nonlinear models of interaction* (something that was also clear from the literature review on this topic). The model also needs to focus on interaction rather than on how other entities—such as computers and users—are interrelated in their interaction. In short, it needs to foreground the entity of *interaction*, and it needs to serve as a bridge between the enabling technologies and materials for inter-action, on the one hand, and interaction from the viewpoint of how these materials are used (during sessions, or as I prefer to call them in this model, during "threads of interaction"), on the other.

This means that the basic model of the materiality of interaction that I present here is a model and a first draft of a theory that hopefully moves beyond the "little man computer" model, and beyond any linear or turn-taking models of interaction that tend to acknowledge the two entities of "users" and "computers" rather than interaction. On the contrary, the model I propose acknowledges how interaction is enabled by these other entities (including users, computers, and other enabling materials); how those materials are interwoven in threads of interactions; how a particular form of interaction can be enabled only given a particular materiality; and how the interaction with and via this materiality also, and dynamically, further forms the materiality of interaction.

One of my goals with this model is to make visible what is otherwise hard to visualize. In interaction design projects, the focus is typically on the design of user interfaces—as a visual manifestation of an interaction design—and on the analysis of user needs, and the development of use scenarios—as a way of highlighting the user as an important entity in interaction design. While such entities—including how interfaces are manifested and descriptions of its intended users—are certainly relevant and interesting, they do not say much about the aspects of interactivity I introduced in chapter 5 (including the pace, rhythm, balance, and responsiveness of the interaction being designed). In fact, while highlighting the entities that come together—people and interactive systems—they miss one core aspect of interaction design, which is the form aspects of the interaction. While typical models and approaches to interaction design put the spotlight on users and computers as enabling entities for interaction, they give very little guidance for reflecting on the materiality and the form of the interaction being designed.

As a response to this need, I will now propose a first draft of a model that serves as an attempt to contribute to these existing models—not as an alternative, but rather as a complement for thinking about the materiality and form of interaction. As such, this proposed model both serves the purpose of enabling discussions and reflections on interaction, and works as a tool that can be helpful when designing interaction.

To return to the two notions of *threads of interaction* (use) and *threads of computing* (processing), I propose that these two processes are fundamental for understanding the materiality of interaction. As figure 6.5 illustrates, not only are these two processes entangled with each other, but it is through this double-loop entanglement that the materiality of interaction is formed. Let me elaborate on this.

A number of fundamental elements form the basic building blocks of this model. As I will describe below, it is through the specific ways these elements are entangled that the form and materiality of interaction are produced, setting the frame for how they can change, and affecting how we perceive the way the interaction unfolds—at the current moment and over time.[16]

So let us start with this notion of *threads of interaction*. In short, a "thread of interaction" is what we typically refer to as the "use" of an interactive artifact. However, while "use" only seems to mean that an object or artifact is "put to use" or "applied" as a tool for doing a particular task, without being affected or changed through its "use," the "thread of interaction" connects the "user" with the materiality being used. A "thread of interaction" is enabled by the materiality of the interactive artifact, and a "thread

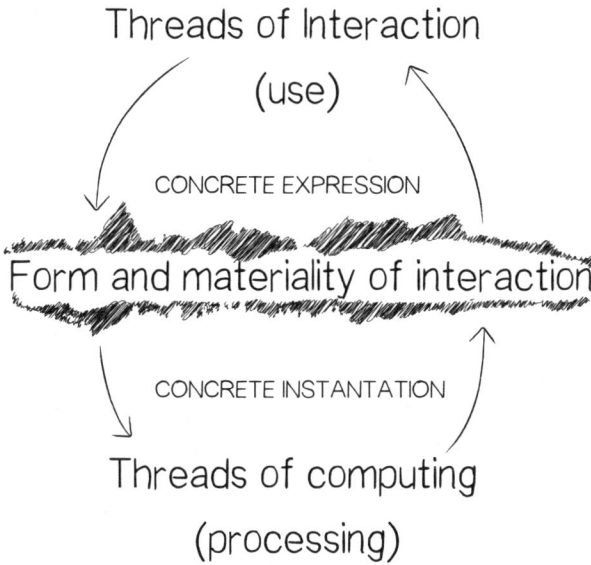

Threads of Interaction

(use)

CONCRETE EXPRESSION

Form and materiality of interaction

CONCRETE INSTANTATION

Threads of computing

(processing)

Figure 6.5
A basic model of *the form and materiality of interaction* in relation to the "threads of interaction" (use) and "threads of computing" (processing).

of interaction" also adds to, and changes, this materiality. As such, the materiality partly emerges and changes during these threads of interaction. As an example, consider a word processor. During "threads of interaction," the word processor is simultaneously a tool and the container for the text written or edited. It is through the threads of interaction with the word processor that text is generated or further edited.

In this model, I have also introduced the notion of "concrete expression" as an element. Typically, we refer to this as the user interface of an interactive artifact: the "concrete expression" is how the materiality of interaction manifests and presents itself to the user. It might take the form of output on a screen, but it can also be manifested in many other forms. Janlert and Stolterman (2017) introduced the notion of *surface-bound* versus *surface-free interfaces* as a way of distinguishing between different forms and manifestations of user interfaces. In short, the concrete expression indicates the form in which the materiality of interaction is concretely expressed. How the materiality of interaction is concretely manifested to the user both enables particular modes of interaction and affects how the user understands and engages with the materials through *threads of interaction*. As a simple example, think of how the surface-bound interface of a word processor enables many visual modes of human-text interaction, whereas the surface-free interface of Siri[17] enables iPhone users to speak to and even converse with their smartphones (a mode of interaction that can be handy when riding a bike, for instance).

It should also be noticed here that since interaction is not a state but rather an ongoing and dynamic process, the "concrete expression" might change over time. For instance, it could change to allow the interaction to continue over a multitude of platforms, applications, devices, and/or other forms of instantiations, or by allowing the expression to shift in shape.[18] Through interaction, new materials are typically added or produced that add to the further thread of interaction. For example, our interaction with a word processor typically also results in text being produced that, in turn, is a material that might be part of further interactions (such as through the act of editing the text).[19]

In parallel to the "threads of interaction" that the user engages in when interacting with the "concrete expression" of the form and materiality of interaction, we also have processes of computation that continuously change the materiality of interaction (and thus change what can be concretely expressed). These "threads of computing" take care of user input to the interactive system, monitor the interaction with the system, and process and store data and information. The threads of computing in this model, together with the set of materials that are configured to be part of the interactive artifact, are what I refer to as the *concrete instantiation* of the materiality of interaction. Here, the "concrete instantiation" is the blueprint of the interaction in computational form at a given moment. This set of materials refers to any material that is configured to be part of the interactive system (including not only digital and physical materials but also, for instance, positions, pace, steps, location, weight, light conditions, and so on). In short, a "material" is in this context anything that a microprocessor can read, write to, register, sense, monitor, or process.

When an interactive artifact that is guided by the concrete expressions of the materiality of interaction is used, the materiality of the interaction typically presents itself to the user in the form of a user interface—whether visual, audio-based, or tangible. Its use also gives form to the interaction and thus affects its materiality, as well as serving as input to the *threads of computing* that in turn also affect the materiality of interaction.

From this model, we can now see how the materiality of interaction might change over time, depending on both the threads of interaction and the threads of computing. In fact, a core character of the materiality of interaction is its ever-changing state and form, its dynamics, and how it performs—both in relation to its user and in relation to the composition that defines the concrete instantiation of the interaction in computational form. We can also see that from an interaction design perspective it is possible to imagine different types of entanglement of the two processes of threads of interaction and threads of computing. For instance, one of the many interaction models possible to imagine and design might include a completely balanced turn-taking (where each turn to the thread of interaction is followed by a turn to the thread of computing, as in traditional command line interfaces). Another alternative might be an agent-based interaction model based on a different balance and thus on a different materiality of the interaction. And yet another interaction model might be one that is focused mostly on "monitoring," a model of interaction that might suit some particular use scenarios (for instance, going for a run with the Runkeeper app operating on a smartphone, as a solution for monitoring one's exercise).

Beyond serving as a concrete tool for imaging different interaction models, this proposed model of the materiality of interaction also illustrates why designing interaction is so complex. I would say that it is complex because interaction sits right in between typical entities of analysis (that is, "users" and "computing systems"), and designers are typically trained to design *entities* (and not the void between entities). The materiality of interaction is also a complex object of design both because it expresses itself through user interfaces and because it is partly formed by the threads of interaction that the materiality enables. It is this duality, its changing form during use, and the interplay between its instantiations and its expressions that make interaction a tricky and complex unit of design.

Overall, the model that I have presented here relies on a theoretical ground that basically says that the materiality of interaction is constantly formed and transformed by three intertwined processes. These three processes include (1) processes of interaction (*threads of interaction*), (2) processes of computing (*threads of computing*), and (3) *the materiality of interaction* as an ever-changing process that reflects the entanglement of the other two processes. As such, this model is not a linear model of interaction, and it is also a step away from any entity-based model of interaction that focuses on the entities of users, computers, and interaction as only being about turn-taking between those two entities.

With this model in place, and with a better, more precise understanding of *the materiality of interaction*, we can now start to consider the model's implications in terms of how it opens up a new interaction design agenda, and to identify some of the challenges ahead of us. The next chapter is an attempt to start addressing those topics. In particular, I will set out to analyze and describe the implications for interaction design as we move from representation-driven HCI, through the material turn, to a material-centered interaction design approach. In doing so, I will in particular focus on how an interaction design paradigm that foregrounds the materiality of interaction at the same time foregrounds (1) the importance of designing meaningful interaction design, (2) the importance of designing "stuff that works," and (3) the importance of authenticity when designing interaction.

As I will illustrate in the next chapter, the move toward a material-centered approach to interaction design has closed the gap between the world of atoms and the world of bits. However, it has also opened up space for thinking about a new materiality, that is, *the materiality of interaction*.

7 A New Materiality
The Rise of a New Interaction Design Agenda and
Three Challenges ahead of Us

The previous chapter provided us with a theory of *the materiality of interaction*. I suggest that we might now need to look back in order to see the road ahead of us more clearly. In this chapter, I will therefore show how a focus on *the materiality of interaction* on the one hand leaves any distinctions between the physical and the digital behind, and how on the other hand it presents us with three distinct challenges as we move forward through the material turn. So let me walk you through this turn—from our background in a representation-driven approach to interaction design and toward this new materiality.

In the first chapter of this book, I discussed how the traditional *representation-driven approach* to interaction design was to a large extent about *the re-representation of aspects of the world in the computer, and on the digital screen.* In one such context, the practice of doing good interaction design was about (1) *identifying important aspects of the world* (for instance, a particular arrangement of an information flow, the sequence of activities, or the states of objects) and (2) *finding a suitable way of representing these different states or aspects* in the computer (in the form of variables, database structures, table identifiers, and so on). To enable interaction with these representations, designers typically have relied on the design of visible/graphical representations, and they have relied on symbols and metaphors to establish and maintain meaningful connections between the representation (in the computer) and whatever it represented (in the world), or they have used symbols and metaphors that rely on a connection or meaning in the physical world (for instance, the "trash can" symbol and metaphor on desktop computers).

Furthermore, to allow for interaction with these representations, interaction designers have invented or developed techniques for interacting with the representations—for instance, through the development of a wide range of different input/output peripherals and interaction modalities (ranging from traditional mouse/keyboard interaction modalities to eye-tracking, gesture-based interaction modalities and user interfaces for importing, moving, erasing, editing, or exporting data and information). As such, the representation-driven account simultaneously came with an idea of computing as being heavily focused on information manipulation, processing, storing, and distribution.

This process of bringing aspects of the word into the computer for further information processing allowed for easy access to information and the construction of huge databases; it enabled information visualization and large-scale storing of data, which later gave rise to today's cloud-based services and big-data initiatives. This focus on information storage, distribution, and easy access took us all the way from early solutions for doing distributed or networked computing to the online streaming services of today.

The representation-driven account has indeed been, and still is, a powerful approach to interaction design. It allows for overviews of data and information, remote collaborations via representations, and a multitude of ways to visualize, explore, and work with data. To enable successful interaction design within this design paradigm, separating the representation from what it represents has been key.

In a way, the stereotypical factory control room is a great illustration of the advantages of the representation-driven approach to interaction design. Instead of requiring people to walk around in the factory and check the machines and processes in the building, the representation-driven account enabled the creation of "bird's-eye" models of the factory and allowed for remote monitoring. This was enabled through the development of large computer screens (or the configuration of multiple screens into "screen-scapes"), computer networks, and the design and development of sensor technologies that could sense and collect data of importance for the overview of the factory.

To find appropriate ways of representing the data on the screens, and to also define and implement alarms that the sensor data could trigger, designers had to develop many definitions of "normal states" or "normal ranges"

Figure 7.1
Illustration of how screen-based visual representations (of some aspects of the world) enable overviews and easy information processing in the context of a factory control room.

and of deviations from these norms, in order to provide mechanisms for triggering alarms in these representational overviews of the factory. Once this accurate representation of the current state of the factory was enabled on the screen, the practice of monitoring the factory could transfer from a person walking the factory floor to a person sitting in the control room; and it further enabled remote diagnostics, storing of process data, logs (to check and analyze the production process), and so on.

In short, the focus on interaction design solutions for working with representations has indeed been a successful endeavor, and it has been the leading design paradigm for the digitalization of many businesses, organizations, and manufacturing processes.

So representations are powerful. A representation allows selected dimensions of the world to be presented in a preferable format. As such, representations are also political, rhetorical, or at the very least somewhat biased: someone has decided what should be shown, and in what format. In addition, a representation can be manipulated without changing the thing, state, or process that it represents. As such, the very definition of a representation is that it holds the presentation of something apart from what it represents. Representations enable the elaboration of different scenarios without actually changing the nature of things. They make it possible to experiment with data, and allow for simulations. For these purposes, the representation-driven approach to interaction design is still valid and strong.

Due to the representation-driven account, we have become accustomed to a processed and mediated understanding of data. A visualization of data is, by definition, also data processed through the lens of a particular form of visualization. A diagram, for instance, presents data in one way, whereas a table might let us interpret data in a different way. But it is not only through different data presentation techniques that we have learned to work with data in a mediated way; our frequent use of symbols and metaphors in interaction design has introduced an additional layer of interpretation, so we need to understand the meaning of the symbol or the metaphor being applied in order to understand what the presentation represents.

In this sense, symbols and metaphors work as shortcuts for understanding, interpreting, and interacting with representations. As such, they operate as mediating structures and interpretive lenses that enable us to quickly make sense of representations and what they represent (including their connection to the underlying data, and then the relation between that data and the world from which it has been collected or generated).

The material turn in HCI had a fundamental impact on this line of development. While the obvious sign was the move from a single focus on the design of graphical user interfaces to alternative ways of interacting with computers—including the design of tangible user interfaces—I would say that that was only a cosmetic move in comparison to the more fundamental shift this turn implied. In fact, the fundamental turn here was not so much a turn from the visual to the tangible, or from the screen-based surface to the handheld object; instead, this turn marked a shift from keeping things apart toward holding things together. Let me elaborate on this.

The power behind the representation-driven approach to interaction design comes from the production of representations of the world—and in a format that a computer can both process and produce. This distinction between the world and the computer is important to uphold for the representation-driven approach to succeed. The representation gains its power from being just that, a representation, rather than the real thing.

On this level of analysis, the material turn in HCI was not a turn away from "painted bits"—that is, "pixels"—toward more tangible ways of representing data; rather, it was about eliminating any distinction between the computer and the world that it once was designed to process representations of. Far beyond any such distinction between entities, the material turn implied that computing should be regarded as an entity in the world, and as part of any other configuration of our everyday life.

So, while the current interest in the Internet of Things (IoT) could be seen as yet another way of giving the computer some kind of more tangible user interface, and while the IoT seems to point toward computing in the form of objects rather than in the form of gray boxes, it is on a more fundamental level related to the way the IoT movement enables the reactivation of everyday materials and things that already surround us. This turn in the development of computing puts the computer "out there," all around us, and in relation to the things, vehicles, objects, and buildings that, taken together, constitute our surroundings. As such, the material turn was ultimately a turn toward a relational approach to computing (in relation to the things that surround us), and accordingly a call for compositional rather than representational interaction design thinking.

One of my main motivations in writing this book was to address this growing interest in the notion of "materiality" in the field of human-computer interaction/interaction design. Certainly I could have written a book about tangible user interfaces and I could have focused on how the design of user interfaces is increasingly about the design of a physical, rather than a digital, user interface for interaction with computers. However, that is only one aspect of the materiality we are dealing with here. Far beyond this literal approach of following the development from the graphical and mouse/keyboard setups for interaction, we need to redirect our focus from the materiality of the digital artifact to the materiality of the interaction. Let me give a simple example to illustrate my point.

Consider the Nike+ sensor. The Nike+ sensor was developed as a wearable and wireless sensor that you could put in your running shoe, and you could then connect this sensor to your smartphone and download the Nike+ app. This configuration enabled the wearer to monitor the number of steps taken and so on. The Nike+ sensor was in that sense a material part of the interaction design, but it also demanded that you physically place this sensor in your shoe as part of setting up this particular interactive system. Far more interesting from an interaction design perspective, I think, is that this configuration allowed the activity of "running," with these particular shoes on, to constitute a new materiality of interaction. Without these shoes on, and without the sensor in the shoe, a run would just be a run; but with these sensor-equipped shoes, the run is more than just a physical activity—it is also a way of interacting with an interactive system, a mode of interaction enabled by a particular materiality (including the smartphone, the sensor-equipped shoes, the Bluetooth stack that creates the link between the sensor and the smartphone, and the software and the digital services that hold this composition together). Accordingly, the materiality of this particular form of interaction is dependent on this material configuration, which is a particular composition that holds the smartphone, the running shoe, the sensor, and the app running on the smartphone together, so as to allow the activity of running to simultaneously work as good exercise and as the main modality for interacting with this interactive system. In this example, the computer was literally placed in composition with the activity being supported, and in composition with other physical and digital materials configured to support this particular activity (including the running shoe).

Thus, the notion of "materiality" does indeed involve thinking about the combination of computational and physical materials, but *the materiality of interaction* suggests that a particular activity in the world (such as the activity of "running") might not only be a physical activity, but if viewed as a form of interaction, it might also be scaffolded by a particular composition (which, of course, includes computational materials).

In this book I have tried to highlight that this idea has fundamental implications for interaction design. Instead of thinking about interaction as (1) being restrained to some form of explicit turn-taking between a person and a computer[1] (whether the computer is in the form of a laptop, a tablet, or a smartphone), and (2) mainly screen/keyboard/mouse-based, we have seen that the last three decades of developments in the area of interaction design suggest that interaction can nowadays take just about any form and be part of just about any activity, and that the material configurations that enable this to work as a mode of interaction can also come in just about any material form. This shift definitely calls for new approaches to interaction design, and it broadens the design space for interaction design—beyond any form of "user-box" interaction model, and toward thinking about interaction in connection to almost any relation across "activity-material" compositions.

The example of the Nike+ sensor illustrates that a key for doing successful material-centered interaction design can stand in direct contrast to the representation-driven approach (both approaches are complementary and necessary, but have different purposes for digitalizing our world). So, instead of keeping things apart, as the representation-driven approach does, a key dimension of doing successful material-centered interaction design is to never introduce any distinctions at all between the computer and the world in which it is embedded, nor to introduce any distinction between the activity of "interaction" and any other activity in our everyday world. Instead, computing is considered a material among other design materials, interaction is an activity among other activities, and sometimes even a seemingly different activity (like the activity of "running") can simultaneously be about interaction. What follows as an implication from this is that good material-centered interaction design is about composing interaction across different materials and activities rather than trying to represent any of it in the computer, or trying to keep any of these things and activities apart.

As an example, the design and construction of a modern car is, to a large extent, based on a complex but still very particular configuration of a number of integrated, moving mechanical parts. As such, a modern car demonstrates compositional thinking in the context of mechanical engineering. But increasingly, a modern car is also to a large extent a computing system. Any modern car is equipped with a number of computers and sensors, but far beyond that technical infrastructure, the mechanical and the computational parts that constitute a modern car are also integrated and intertwined. This tight integration of computational and mechanical parts not only allows us to label a vehicle as a "modern car," but to talk about *the materiality of interaction* in relation to the activity of "driving" a modern car. When we drive a modern car, our way of driving is reflected on the dashboard. We might, for instance, get indications if our style of driving falls in the category of "eco-driving," and some cars even allow for different configurations of the driving experience through the push of a button. By pushing a button, the driver can switch between "eco," "normal," or "sports" modes of driving, and each push on the button is also reflected in different reconfigurations of particular (physical) properties of the car. For instance, if the driver selects the "sports mode," the car changes its character to allow for faster accelerations, harder braking, and more direct steering. By allowing the digital infrastructure of the car to work in composition with the physical and mechanical parts of the car, the materiality and thus the experience of "driving" as the main form of interaction with the car fundamentally change.

But this development is only the beginning of how the material turn will change the experience of "driving" a car. Increasingly, the car industry is exploring solutions for so-called self-driving cars. When self-driving cars are in traffic, the activity of driving will be a matter of selecting where to go, monitoring the car while traveling, and experiencing one's surroundings during the ride.

As the activity changes, so does the materiality that both enables and influences that activity. As figure 7.2 illustrates, this will imply very obvious and observable changes in how the driver will interact with the car (for instance, there is no need to keep the hands on the steering wheel while driving) but will also affect what materiality is activated while driving (such as GPS technology for reaching a selected destination, collision detection technology to enable the car to detect and avoid obstacles, and so on).

On a more general level, the example of the self-driving car, with the changing mode of interaction from the driver's perspective (and also for passengers and other people on the street) and a new materiality of interaction at play, also opens up a whole range of additional questions about the future of interaction design as a design discipline, and how we should think about the broader design space for interaction design. In short, what *is* an interaction design problem? And what problems can be solved with (good) interaction design? Is interaction design only about "active interaction"? In other words, do we only count active interaction as an object for interaction design? And in the case of the driverless car, is the user still "active" when the car is moving? If so, in what sense? By monitoring the car? Or is "traveling" a different way of interacting with the self-driving car? If so, then what is interaction design in this particular context?

Furthermore, can we think about different levels of engagement in interaction design? For instance, one could be in a highly "active mode" of interaction when telling the self-driving car where to go, and then be less active, maybe even asleep, while the car travels to the preferred destination?

Figure 7.2
The interior of a self-driving car where the driver can rest her hands on her knees instead of holding the steering wheel while driving.

To return to a more general question: Can we think about interaction design as a case in which we move in and out of the bubble of interaction (in and out in terms of our engagement with the "threads of interaction") while the computer, in this case manifested in the form of a car, continues to process the necessary "threads of computing" in order to take us safely to our selected destination? This idea goes along with the notions about interaction that were introduced with the model of the materiality of interaction toward the end of the last chapter.

In this example, the computer is manifested in the form of a car and is part of a computational system (including the roads, satellites, surrounding traffic, etc.), but it's also part of a collaborative social system (including other drivers, pedestrians, and so on), and of course also part of a legal system (which allows for self-driving cars in traffic and requires the self-driving car to operate in relation to the formulated traffic rules and laws).

Both the example of the self-driving car and the previous example of the Nike+ sensor illustrate computing that is contextualized as part of the world and part of everyday activities. However, the self-driving car also illustrates that the representation-driven paradigm and the material-centered paradigm are indeed complementary approaches that need to work in concert.

For instance, for the self-driving car to operate in the physical world (both as part of the world and as a computer manifested in the form of a car that actually moves around in the physical and social world), it still needs to rely on a good understanding of that world; that is, it needs to rely on good representations of that world. The self-driving car can achieve this representational understanding of the world through the processing of multiple data sources including sensor data, positioning data, etc. Beyond the representational understanding of the world, however, the operations that the computer performs—in the same world that it is simultaneously creating representations of—and the way it behaves and acts as an actor in the world, and in relation to other actors in that world (for instance, other vehicles, obstacles, pedestrians, and so on), pose a question for interaction design that sees computational objects (including self-driving cars) as part of the composition being designed.

So here is a crucial aspect of the material turn that is important to remember: the material turn was never a linear turn *away from something* and *toward something else*, but rather a turn in the history of HCI design that opened up the possibility for new and additional ways of doing interaction design. A major implication to keep in mind is that *the material-centered design approach is not a radical alternative to the representation-driven approach, but rather a necessary complement.* More specifically, it is a complement that highlights aspects of how the computer presents and manifests itself in the physical and social world, in relation to other materials and entities, and in relation to social activities and behaviors. Furthermore, it highlights that we can only arrive at good interaction design if we can manage to work across these two approaches. And to do so, compositional thinking might be a key factor.

Clearly, the combination of representational and material-centered approaches to interaction design enables us to push the development of interaction design forward, but there are also, as stated at the beginning of this chapter, some specific challenges ahead of us. While "the turn to the material" is no longer a challenge or an obstacle, but rather a precondition for how interaction design is developing at the current moment, we now need to consider what other challenges we will face as we move forward. In the following sections I will sketch out and reflect upon three of those challenges.

To identify these three challenges, I have taken the following factors as points of departure: (1) *the materiality of interaction* as a new ontology for interaction design, (2) *material-centered design* as a complementary approach to interaction design, and (3) *manifestations* (rather than representations) as the metaphysics of interaction design. In relation to these three factors, I notice that the design of new materialities is not only about making real, but also about *making meaningful and making sense*, that material-centered design is ultimately about *making stuff work*, and that manifestations are all about *authenticity* (whereas representations can be measured along the scale of how well they represent something else, independent of how authentic or inauthentic that other thing is). In the following sections, I will sketch out how each of these three challenges unfolds in relation to a design approach to the materiality of interaction.

Challenge 1—"Making sense": On meaningful material interaction design

If *the materiality of interaction* is constituted through the entanglement of *threads of interaction* and *threads of computing*,[2] where the threads of interaction are intertwined with our everyday activities and the threads of computing work across many different materials and substrates, then we need to make sure when doing interaction design that these arrangements of activities, materials, and computational resources are not only usable but also meaningful. When I say "meaningful," I mean that there needs to be a meaning, a purpose, a clear idea, or a well-defined concept behind the design,[3] and hopefully, the user of the design will be able to understand and appreciate that meaning. For instance, a sweater is not only the arrangement of wool into a larger structure and, as such, the end result of many hours of knitting as an activity, but it is also designed to keep you warm, and hopefully to make you look good when wearing it. The yarn has accordingly been selected and arranged in a meaningful way.

If one is aiming to conduct meaningful and material-centered interaction design, there are a few aspects to consider. First of all, material-centered interaction design poses the challenges of making meaningful choices of materials and of devising meaningful ways of bringing different materials together. Both challenges involve the meaningful use of materials in the design.

In addition, the design needs to be meaningful. If there is a disconnect between what a user wants to do and what the digital service or device does or offers, the design will be perceived as less meaningful. If it is hard to understand the design, it will again be perceived as less meaningful. Accordingly, meaning is closely related to concrete functionality as well as perceived functionality. In short, meaning is related to how a device works, and in the context of interaction design, it is about how the form of interaction is crafted and manifested in material form. Meaningful interaction design thus becomes a design process aimed at (1) aligning the form of interaction with the activity it is intended to support, and (2) making sure that the form of interaction being designed also works in composition with the materiality that enables and supports that particular form of interaction. Finally, I suggest that (3) meaningful design of interaction also deals with the design of meaningful flows of interaction—that is, the "threads of interaction" are designed in a way that makes sense, are aligned with the activity being supported, and unfold in a clear and meaningful way in relation to that particular activity.

What is understood as meaningful is also defined by the person using an interactive system, based on his or her preferences, knowledge, and expectations. Novice users might, for instance, interpret a solution in a different way than an expert user. And while a novice user might need more clues about how the system works, it is typically the case that expert users want very clean and fast interaction models. These clues for novice users might take the form of explicit instructions or help guides, or might be more implicit, communicated in the form of the affordances of the interactive system.[4]

While "affordances" can communicate the meaning of an interactive system to its user in the form of how the user perceives the system's functionality, there are many other dimensions to this notion of "meaning." How we interpret meaning and what we find meaningful are relational, and are highly influenced by a great number of different factors. For instance, the context we live in (including our physical, social, historical, symbolic, cultural, ethnical, and religious context) does indeed affect how we see and ascribe meaning to what we perceive. Culture is a process, and so is sensemaking. Our understanding of things changes over time, as do our values and what we find, experience, and value as something meaningful.

We can also think about this notion of "meaning" in relation to material-centered interaction design from the viewpoint of Bruno Latour's description of the relation between material configurations, everyday activities, and the way our material surroundings affect and regulate human activities.[5] As an illustration of this, let us look at the classic example from Latour (2007) of a "speed bump" as being not only a physical obstacle on a road to regulate traffic, but also a phrase with many social inscriptions.

Latour (2007) explains that a speed bump is a material manifestation of social rules, and simultaneously is a material manifestation that has some agency. The design of the speed bump is deliberately formed to force cars to slow down in order to pass over the speed bump in a comfortable way. If a law or a rule has been formulated that states that we should slow down on a particular road, then we can also physically manifest that rule by placing a speed bump on that road. As a material instantiation of this rule, the speed bump does a couple of things. First of all, it is a very visual sign that tells the drivers on the road to slow down. Secondly, the design of the speed bump actually forces drivers to slow down (if they don't want to have a very bumpy ride or, in the worst case, to damage their car). Thus, the speed bump effectively affects the behavior of drivers as they approach it. For sure, we can think about these speed bumps as "sleeping policemen" because their particular material configuration acts as if it were a policeman telling the drivers to slow down. Further, this example shows that there seems to be a strong connection between meaningful configurations of materials and how those configurations stand in relation to preferred ways of interacting (in this case, how the driver "interacts" with the road).

On a more general level, we can say that the materiality of this law—to drive at low speed—is manifested in the physical form of the speed bump. While this materiality regulates the pace of traffic, the car's suspension also mediates the impact of the speed bump, from the viewpoint of the individual driver. As a result, the materiality of the speed bump and the car's suspension mediate the interaction between the driver and the road.

Indeed, an explicit challenge for good and meaningful interaction design is to design the materiality of interaction so that it reflects the designer's intention of how the materiality should enable or enrich a certain activity or experience, while also (at least in some cases) restricting or even preventing other activities from happening or being undertaken.

Challenge 2—"Making it work": On material interactions and integrations

To do interaction design with a particular focus on the design of *the material-ity of interaction* means to configure the material so as to support a particular form of interaction. Here the ultimate challenge is to make sure that there is a good fit between (1) the *intended form of interaction* being designed, and (2) the design of the *material configurations* that should support or enable that particular form of interaction. If there is a mismatch between these two, then there is a great chance that the design will fail. In short, the challenge is to make the composition work.[6]

To make the composition work is a relational concern for interaction design. Whether it "works" is a matter of whether the material configuration works in relation to the activity it is intended to support. Furthermore, whether it works as intended is a matter of whether or not the materiality instantiates the intended form of interaction that the interaction designer had in mind. In short, it depends on whether the particular form of interaction, as imagined by the interaction designer, did emerge in the interplay between the user and the material composition.

Clearly, this notion of "working" is not only about getting the various sensors to work in composition with some computational resources and some other additional materials; it is also about validating whether the assumptions made about the design are correct. For instance, think about the use of embedded rain sensors in the windshield of a car. When such sensors are embedded, the car can "sense" if it is raining (or if any type of fluid covers the windshield), but the main question from an interaction design perspective is what that sensor data should activate. For instance, should it only warn the driver about the current condition (water on the windshield), or should it turn the wipers on/off? And should the user/driver be allowed to interfere? If so, how? This goes back to the question of how to balance the interaction between the human and the machine. As seen from this simple example, the question of just getting the device to work in a good and proper way quickly leads to a number of additional, but still crucial, questions from an interaction design perspective.

My general advice to address this challenge is to focus on how the composition (rather than simply the user) makes sure that the design works. This is a central concern when doing interaction design with a focus on the materiality of interaction. From the user's perspective, the user should be able to focus on the computational whole and the interaction he or she wants to engage in. Following from that, it is the interaction designer's responsibility to make sure that the user does not get stuck with getting the things to work. Unfortunately, this happens too often; for instance, people get stuck with syncing their devices, trying to connect to wireless networks, trying to pair their devices over Bluetooth, are asked to update or forced to reboot their computer during a public presentation, and so on. From an interaction design perspective, making the user responsible for getting an interaction to work is unacceptable. Good interaction design, which foregrounds "interaction" as the focal entity being designed, should make sure that it is indeed interaction (and not problem solving in order to get the interaction going) that is the design objective. To conclude this challenge, a focus on the materiality of interaction is necessary to make it work.

Challenge 3—"Making it real": On authenticity in material interaction design

A major challenge for doing material-centered interaction design is not only to give the interaction design a material manifestation—to manifest the materiality of interaction in computational and physical form (for instance, in the form of a tangible user interface, a self-driving car, or a smart thermostat)—but to fulfill the related, but still fundamental, relational design challenge of making sure that this material instantiation works in composition with the surrounding material context. That is, making an idea about interaction real involves making sure that the selected material form is perceived as genuine, trustworthy, historically and culturally rooted, and accurate. In short, a major design challenge is to design for *authenticity*.

If we once again take the car industry as an example, we can quite easily see how this challenge presents itself. The car industry has worked explicitly with material selection and ways of communicating value and symbolic meanings via materials for more than a century. Designers in the car industry are aware of what different materials communicate, and how the combination of different materials and forms in turn shape our understanding and interpretations of different cars. A "luxury car" or a "sports car" presents itself as such through the ways in which different materials and forms come together as a composition. The particular ways the materials are configured and compositionally arranged determine how we interpret the vehicle. The materiality of the car as this composition of materials, forms, and color is strongly connected to how we understand a particular car, and what symbolic and cultural values we add to our understanding and interpretation of a particular car.

However, if we look at the technology industry, we realize that it has just recently discovered and started to explore this link between the composition of materials and how we interpret and assign value to things. For instance, when digital technologies are developed for cars, those artifacts might add good functionality, but in terms of material composition there is still a long way to go. Typically, devices like navigators or CD/radio sets for cars still look plastic, and in that way the materiality of these devices doesn´t fit in with the rest of the car and what the car communicates in terms of its overall aesthetics and symbolic value.

As we move more and more toward material-centered interaction design solutions, where the digital is supposed to work in composition with other material objects in our everyday life, we need to avoid solutions that look kitschy or cheap in relation to the material context in which they are supposed to work. This is a relational concern and challenge for interaction designers who decide to focus on doing material-centered interaction design.

But what kind of guidance can we find for moving forward? Are there others who have found solutions to this design challenge? There are indeed a number of related areas of design that we can learn from. I have already mentioned the car industry, but we can also learn from the fields of fashion and architecture. These areas have for centuries considered material compositions in relation to human activities and have developed a good understanding of what different material compositions might communicate in terms of symbolic and cultural values and meanings.

The challenge of "making it real" is also a challenge to decide exactly how, and by which (material) means, interaction should mark its place in our everyday world. Sometimes we see "Scotch tape solutions," so-called because such solutions barely work. In other cases, the solution works so well that we almost think of it as "magic"—it just works, but how it works has been elegantly hidden in the overall design. On a general level, we can say this is the difference between terrible and elegant solutions, distinguishing between compositions that do not really work and compositions that work so elegantly that the parts are almost impossible to distinguish from the whole. Elegant solutions not only present us with working solutions but offer us the opportunity to focus on the activity or experience. As such, good material integration goes beyond the first step in design of merely enabling a particular functionality. Elegant compositions place a particular solution in the world in such a way that the solution also becomes fully integrated in the world and "disappears." The design becomes part of what is real.

The challenge of "making it real" is about designing well-working and authentic solutions, but this challenge is also, and maybe ultimately, about designing long(er)-lasting solutions. Sustainability is an increasing concern in the area of interaction design,[7] and if we set authenticity not just as a design challenge but as a design goal to reach, then to strive for a "good fit" in design projects implies both that the final design works in composition with a larger material and social context, and that it works as a longer-lasting solution.

The challenge of "making it real" is thus a challenge to design something good—something sustainable in its environmental footprint, but also sustainable from a social, cultural, and ethical viewpoint in terms of long-lasting design. Another challenge is to ensure that the design works as a connection between the user and the materiality that enables a particular form of interaction. This connection, however, demands not only that the material configuration is sound, that it is *authentic*, but also that the user commit himself or herself to using the interactive system in the intended way.

If we take "eco-driving" functionality as an example, we can think about it from a purely functional perspective, for instance in modern cars where the driver can select an "eco" mode. However, such functionality is useless if the driver does not practice eco-driving by avoiding fast acceleration, high speed, and hard braking, which all lead to higher fuel consumption.

For interaction design, "making it real" is thus a challenge of designing materialities for *preferred forms of interaction* that users are also willing to subscribe to. As we move forward, we will see that this is a fundamental challenge in taking a point of departure in the material turn as the first step toward the making of a new real. The concluding chapter of this book is therefore devoted to reflections and speculations, and to my thoughts on where we might go from here. In short, I will reflect on what the implications might be if we fully approach interaction design from the viewpoint of material matters.

8   Moving Forward
From Materials to the Making of a New Real

At last everything must come to an end. Instead of concluding this book by summarizing some of its main arguments, I see this final chapter as an opportunity to look back and also to think about how to move forward. The focus of this book has been on the "material turn" in our field—how it has played out historically across the history of computing, how it has fueled a discussion of materiality in human-computer interaction, and how it now sets the scene for a material-centered approach to interaction design. With such a focus, I think that it's important to understand that a *turn* is not a state but rather a movement, and thus not just a process we should follow, understand, and describe, but something that we should also probably try to trace the origins of and try to make some predictions about in terms of where we are going. In short, how did we end up here? And where are we going next? In line with these ideas, I suggest that it is now time for us to look back, in order to move forward.

I started this book with a historical backdrop to illustrate how interaction design has, during different phases of our history, always been related to *materials*, although in different ways—from the early days of computing, when paper-based punch cards were essential materials for representing and processing data, to the tangible computing objects of today.

Across the history of human-computer interaction, the role of materials in relation to computing has changed. The paper-based punch cards were used to represent data, and accordingly worked as a physical way of storing data for the computer to process. Today, this way of using physical materials to store data is still valid (just think about the technical and material details of how any hard drive works), but now materials are used not only to store and *represent* data but also to *present* computing in material form.

The movement toward the "Internet of Things" might be one of the most obvious signs of how that line of development plays out in practice.

One way of describing the history of interaction design, and the role of materials in that area, is to say that *materials have always been essential to interaction design*. Another way of describing this history is to say that its relation to materials has shifted over the years. The introduction to this book started with a description of the representation-driven approach to interaction design where materials were used to *represent* aspects of the world. This approach contrasts with the contemporary Internet of Things movement, in which materials are used to manifest the computer as part of the world. This shift, from representing aspects of the world to partly manifesting the world through the use of (computational) materials, is one way of describing the change in relation to materials in interaction design.

As we focus on this shift, we can also think about it in relation to different trends. Of course design trends come and go, so one might ask how we should understand the current moment as well as our past (as understood through the lens of the current moment). If we stop and reflect for a moment, can we see the past, the present, and the future with clear, objective eyes? For sure, design trends also reflect the current design paradigm, and it is within the realm of the contemporary that we need to look at how and why things are developed in a certain way. For instance, the trend toward skeuomorphic design occurred at a time when there was a strong need to help people move from the physical world to something similar that computational devices could offer (such as e-books, digital notepads, and so on). Borrowing the look and feel of the physical book and the paper-based notepad was meant to make the new design feel comfortable and familiar—and indeed the design trend toward skeuomorphism did help to ease the transition in moving from the physical to the digital.

As I have discussed in this book, this design trend does, however, by definition,[1] uphold a clear distinction between the physical (the real) and the digital (the representational). The skeuomorphic design of an e-book strives to make it look like a physical book, but something is lacking; the artifact might resemble a physical book, but it will never be the real thing. In fact, this movement toward the skeuomorphic design of e-books has generated lots of reflections on whether e-books can replace or provide the same experience as physical books. While the pages could look like traditional paper, and while the turning of pages could be animated to look like pages turning in a physical book, the e-book would still not be the same, nor could the design go beyond the physical book because it relied on a metaphorical connection to a similar object in the real (physical) world.

Because skeuomorphic design helps people to understand the digital through the physical, it relies on a metaphorical understanding of the digital (it was "almost" like the physical book, but never was that physical book). Since the digital is only representational, it could, at best, only mirror the real, but never move beyond that metaphorical connection between the physical and the digital.

So, to summarize, we can say that skeuomorphic design relies on (1) a metaphorical connection between the physical and the digital, (2) the idea that the physical artifact has valuable properties, but that the digital does not have any unique properties, and thus, (3) the conclusion that the physical could work as a foundation for digital design.

In an attempt to break away from this asymmetrical design trend, some initiatives were taken in the opposite direction. Non-skeuomorphic design was introduced as an approach that suggested that digital design does not need to rely on any heritage from the physical world. Instead, the digital has its own properties, and thus digital design could be done without any linkage to the physical world. However, this opposite design trend also came with its own drawbacks. If not inspired by the physical world, then what could serve as the guiding principles for digital design? So-called flat design grew in popularity, with minimalistic user interfaces with no "physicality." Such designs lacked any user interface elements that would remind the user of things in the physical world, such as animations, "buttons" that the user could "press," shadows to indicate "buttons" pressed "down," and so on. In fact, even the word "pressed" sounds like something we do with some force and some friction—qualities of the physical world but not necessary in the digital world.[2]

"Non-skeuomorphism"—One step further away from aligning computing with reality?

From one perspective, the trend toward non-skeuomorphic design did solve a number of user interface design challenges. It suggested that user interface design does not need to rely on any metaphorical connection to the physical world and so could take any form, even a minimalistic, almost nonexistent form, in terms of how the user interface is graphically implemented.

Non-skeuomorphic design also pushed for a perspective on the digital as a "unique" design material that differs from other physical matter. Standing free from any heritage or traces of the physical world, it opens up possibilities for new design solutions. As such, it introduced, although in a different way, both a distinction between the digital and the physical and a rejection of the physical. Accordingly, it introduced a useless position for moving toward material-centered interaction design. Whereas skeuomorphic design failed to acknowledge the digital as a unique design material (and failed to provide a basis for compositional design across the digital and the physical, as the latter could at best "inform" the first), non-skeuomorphic design rejected the physical in an attempt to foreground the digital, again suggesting a design approach that kept the digital and physical apart rather than providing ways to work compositionally across these substrates.

Given these two design trends, each with its own limitations but with the same effect, an alternative design approach was needed that could unite rather than distinguish between different matters. In this book, I have presented *material-centered interaction design* as one such approach. I have argued that (1) any material (physical or digital) can be part of the materiality of interaction, (2) interaction design is about the design of this materiality, and thus (3) an interaction design approach that wants to fully engage with the materiality of interaction needs to address how the digital and the physical can work together in composition.

This material-centered interaction design approach builds on a compositional approach to the materiality of interaction. In doing so, it allows for interaction design projects that seek to explore interaction design beyond "the box"—beyond computing in any traditional form (restricted to only a digital matter)—and beyond computing, or "the digital," as a separate concern. On the contrary, the material-centered approach introduced in this book assumes that computing and interaction are substrates and activities fully integrated in our everyday life—from the tiny object and the cloud services that are always present, to the way "driving a car" might be both the practice of steering a vehicle and the simultaneous mode of interacting with the vehicle.

But beyond all of this, and maybe most importantly, we should acknowledge that this third position of *material-centered interaction design* as proposed in this book (1) does not reject the physical world, (2) does not lean toward either the digital or the physical, and (3) does not uphold any distinction in the first place between the physical and the digital. On the contrary, the material-centered approach to interaction design recognizes that every material (digital or physical) has properties that should be carefully considered in every interaction design project—and in any computational composition. It is when we reach this integrated way of seeing things that we can start seeing the world with new eyes.

A new realism—Or seeing with new eyes?

The real voyage of discovery consists not in seeking new landscapes, but in having new eyes.
—Marcel Proust

It is sometimes claimed that the development of new technologies not only gives us new tools and artifacts but also changes our reality. When new technologies are available, some existing practices are also replaced by them. However, while our reality is changed through the producing of new artifacts, it also changes through our understanding of the contemporary, which in turn also shapes how we approach the world. In order to move forward, as Marcel Proust suggests, we might not (only) need to develop new technologies but also to reimagine the existing ones, through the lens of the new. This book has presented material-centered interaction design as one such lens—a particular perspective and approach to interaction design.

I have proposed material-centered interaction design as a compositional lens that focuses on interaction design projects in which (1) computing can take any material form, (2) interaction design is aimed at being implemented across a number of different materials, and (3) the reimagination of traditionally noncomputational materials is allowed to be part of the possible materialities of interaction. I have given the example of the activity of "running" to show how it is not only about exercising, when examined through a material-centered approach to interaction design, but can be reimagined as a mode and modality for interacting with a set of materials and objects (including running shoes) which, taken together as a composition, define that particular interactive system—both in terms of how the materiality is manifested and how it is expressed.

Furthermore, if we examine the current movement toward the Internet of Things through the perspective offered in this book, we notice that this trend is not so much about pushing the technological envelope—in terms of inventing new technologies and new interactive artifacts—as about seeing how the already existing can be reimagined through a digital lens. In fact, it will be literally impossible to replace every everyday object with something that is similar but connected or "smart." Instead, we need to think about how interaction design can work as an approach for enrolling already existing everyday objects in "threads of interaction," to explore how everyday objects can work as input to "threads of computing," and how the *materiality of interaction* accordingly can be designed in order to unify computational elements with the reactivation of what is already part of our world. As such, the true challenge for any interaction designer working in this area is to reimagine the existing through the lens of material interaction design—to design interaction across existing substrates and objects, and to think compositionally about what additional elements and technologies need to be developed in order to get the whole composition in place.

On a general level, this particular shift in "ways of seeing" has to do with a shift in how we see and interpret our reality. In philosophy, this question of how we see and understand our world is a matter of ontology. According to *Wikipedia*:

Ontology is the philosophical study of the nature of being, becoming, existence or reality as well as the basic categories of being and their relations. Traditionally listed as a part of the major branch of philosophy known as metaphysics, ontology often deals with questions concerning what entities exist or may be said to exist and how such entities may be grouped, related within a hierarchy, and subdivided according to similarities and differences.

If we think about this definition in relation to the challenge of "seeing with new eyes," we realize that the most fundamental challenge here is not the development of the new, but undertaking an *ontological shift* in which the existing is not only discovered but is reimagined as computational resources—where something already existing comes into being in a computational moment. The notion of "interactables" is such a reimagined resource. Basically, I view an "interactable" as any substrate, material, or object that is reimagined through a material lens[3] in order to be part of the interaction design of an interactive system. Furthermore, I believe one of the most central skills needed for doing this form of interaction design project will be this ability to rethink existing materials, activities, and practices from the viewpoint of how those might work as part of the materiality of a particular form of interaction, and accordingly as part of an interactive system. Such systems might stretch across the full spectrum, from well-known materials, practices, and activities to the new and the formation of new compositions, where the parts are to a great extent familiar but where computing enables the parts to work together in radically new ways. Creativity, from the viewpoint of interaction design, will to a large extent be about seeing how the already existing and the traditionally noninteractive can be reimagined as part of something new, something interactive, and something that redefines our understanding of how things work.

Movements: Maker cultures, creativity in interaction design, and the DIY movement

Communities are already forming that explicitly try to reimagine the existing so as to invent new things. The practice of hacking goes way back in the history of computing as a way to tweak interactive systems into new forms and to support a functionality not intended or imagined by the designers. Over the last few years, interest in reimagining what computing can do has been growing in our field. Today we can see a number of ongoing initiatives that illustrate this new way of seeing, and new ways of doing creative interaction design work. Simple programming languages such as processing or simple computing platforms such as Arduino, used by professional programmers, are also rapidly becoming tools for creative work in the hands of artists, designers, and "makers." To "hack" and "make" stuff is no longer restricted to inventors and designers; there is also a societal movement that sees these activities as a way of approaching and exploring the world. Hackathons and Makeathons are now frequently arranged to bring people together to collectively explore what a maker approach to the world can offer, and "making" has quickly established itself as an approach to creatively reimagining our world. This movement is likely to continue and to grow, further fueled by the development of hackerspaces, the establishment of FABlabs, and the DIY (do it yourself) movement.

From the viewpoint presented in this book, this movement is a natural consequence of how computing and interaction design are finding their way together out into the world. As long as computing was a separate concern, there was also a separated group of people who devoted their time, competence, skills, and efforts to this niche. And while computing was a separate concern, the materials were in the hands of computer scientists. Today, the material-centeredness of interaction design and the expectation that interaction design should work as an integrated element (in cars, in buildings, in everyday objects, and in clothes, just to mention a few examples) has led to transdisciplinary approaches to exploring the digital in relation to the physical. This development of tools that allow for the easier integration of physical and digital components in interaction design projects has opened the door for nonprofessionals to explore alternative materialities for interaction design.

Designing interaction—Beyond materials, and toward new materialities

This book has certainly advocated for materials in interaction design. I have discussed the representation-driven approach to interaction design, what the material turn in interaction design implies, and I have suggested material-centered interaction design as an approach to work compositionally across the digital and the physicalWith this foundation for interaction design in place, one might wonder what is just around the corner. Given what we know, what do we need to prepare for as we move forward?

As I have already made clear, computing is increasingly a material concern. From the smart watch to the self-driving car to the Internet of Things, the design of the digital is increasingly and simultaneously about the design of the physical. This trend will indeed continue to grow, and we will certainly continue to push the envelope for this type of technological development—but interaction designers should also be aware of a number of additional and related design trends that are growing at the moment.

Material-centered interaction design typically results in very visible and tangible interactive objects. User interfaces are designed that we can see, touch, bend, lift, and so on. This type of interaction design enables its user to literally get in touch with the interactive system. In this paradigm, interaction almost equals a bodily engagement with the materials. But we have a number of other design trends moving in parallel at the moment. One rapidly growing trend is "faceless interaction,"[4] which is the design of user interfaces that have no visual interface. Instead, the user interface might be an intelligent agent that you talk to (as in the case of "Siri" on iPhones, or the Amazon Echo product). I call this a trend toward *invisible interaction.* Another line of development that moves away from the material is the trend toward cloud-based digital services where the user should not have to bother about exactly where or on which machine or device a particular file is stored. Instead, information and computational resources should be accessible at any time, from anywhere, and from any devices. The central idea of cloud-based services is to free the service from any material boundaries.[5] I call this a trend toward *immaterial interaction.* The development of self-driving cars and robotic lawnmowers can serve as an illustration of yet one more line of contemporary development. These interactive systems are designed to minimize the interaction with the device as much as possible. Instead of needing to be actively driven, the self-driving car can autonomously take you anywhere you would like to go. I call such developments a trend toward *minimalistic interaction.*

If we pull these trends together, we can see an interaction landscape forming around us that will not only be about material interaction but also about invisible, immaterial, and minimalistic interaction. It will be about cloud-based and faceless interaction, to the extent that direct interaction might not be needed. With the development of artificial intelligence and developments in the areas of big data and advanced algorithms, one can imagine a scenario in which interaction design, as the practice of designing direct human-computer interaction, is no longer needed. However, that would again mean restricting our way of seeing and only acknowledging certain design elements of our world. So I propose the following alternative way forward.

Interaction design is about designing interaction through the design of interactive systems. Given that computing is now such a fundamental aspect of our reality and our everyday lives, it is also increasingly about designing how we should relate to everything and everybody around us. As computing becomes part of everything around us, interaction design becomes the area of imagining, exploring, and designing our relations with and through these things. As such, interaction design might not foremost be about the design of direct interaction with things and objects, but increasingly about *implicit interaction* and how we should *relate to* rather than actively and directly interact with such an interactive system. In fact, as we move closer to the materials of our everyday life, the practice of interaction design might be less about the design of material user interfaces and more about a compositional design practice where a focus on interaction models for implicit interaction design helps us to bring together elements of *invisible, immaterial*, and *minimalistic interaction* designs with the design of material computational objects. Just think about the robotic lawnmower. What if the robotic lawnmower is part of an ecology of interactive objects, connected to a cloud service, and accessible from different computational devices: what use scenarios can we imagine, just taking this still-limited interaction ecology as a simple example?

Beyond any such concrete examples, whether related to self-driving cars, robotic lawnmowers, or even scheduled backups to online cloud services, we can certainly design a fully automated everyday life—but do we really want that? Interaction is also about *engaging ourselves* in activities and in our world. When we interact, we pay attention, we decide, we experience, we explore, we learn, and we accomplish things. As human beings, we are active, and although we want technology to make our lives easier, we do not want to disengage ourselves. On the contrary, we want to be active, engaged, and involved. We do not just want to be alive; we want to live and experience our everyday lives. For this reason, it is not only interaction design that is a relational concern, but also how the materialities of interaction that we form stand in relation to the different roles we want technology to play in our everyday lives. To imagine such roles, and to think about interaction design as an act of imagining possible futures, will accordingly be an increasing concern for interaction designers.

In *The Sciences of the Artificial*, Herbert Simon (1969) discussed design as being about the design of *future preferable situations*.[6] In line with Simon, I would suggest that interaction design is about the design of the artificial, and thus interaction designers need to be good at imagining preferable futures. Interaction design is not "given by nature," nor restricted by the laws of the physics,[7] but is something for us to design as part of a future world, where we as interaction designers have the responsibility to imagine "future preferred situations." This responsibility transcends thinking compositionally across physical and digital materials. Far beyond any such endeavor, I would say that interaction design is always and more fundamentally a social and ethical concern. To tackle such fundamental design challenges demands a number of things. It demands a good understanding of *the materiality of interaction* as the object of design, it demands *a compositional approach* for thinking about how different things come together as a whole, and it demands *an informed historical account* in order to move forward. My hope is that this book has contributed to the further development of such an informed and reflective approach to interaction design.

Notes

Introduction

1. For an in-depth discussion of the material turn in HCI/interaction design, see Robles and Wiberg (2010).

2. I should also note here that our field has acknowledged other "turns" in its development. For instance, Krippendorff (2005) addressed the "semantic turn" in our field and how it presented a new foundation for design, whereas Grudin (1990) acknowledged a "turn to the social" with the birth of the research area of computer-supported cooperative work (CSCW).

3. For some related work, see for instance Vallgårda and Redström (2007) on computational composites; Vallgårda and Sokoler (2010a; 2010b) on material properties of computers; Jung and Stolterman (2011) on digital form and materiality; Gross, Bardzell, and Bardzell (2013) on the materiality and medium of interaction; and the most recently published paper on the materiality of media by Dourish and Mazmanian (2013). In chapter 3, I also further explore some of the related work with a particular focus on how our field has theorized interaction in relation to this notion of materiality.

4. For an in-depth discussion of the third wave of HCI see for instance Bødker (2006) and Fällman (2011).

5. See Weiser (1991).

6. See the pioneering work by Hiroshi Ishii et al., including Ishii and Ullmer's seminal paper "Tangible Bits: Toward Seamless Integration of Interfaces between People, Atoms, and Bits" from 1997.

7. See for example the work by Ishii and Ullmer (1997), Ishii (2008), and onward as excellent illustrations of this trend.

8. Chapter 1 of this book provides a historical account of HCI, focusing in particular on the history of representation-driven design in HCI.

9. Think about "check-ins" as one example where our physical location is used as input to a social network service, or take as another example "cloud services," in which we try to hide the exact location of where data is stored. In the first example, something physical (location) is linked to a digital service that can be reached from anywhere, whereas the second example illustrates how an aspect of computing (storage) that historically has been bound to a particular location is now distributed to such an extent that it can be labeled as a "cloud"—accessible from anywhere, and in a form that suggests that even computing no longer needs to be bound to any particular, and physical, machine, hard drive, or location.

Chapter 1

1. The notion of "painted bits" was first coined and introduced by Hiroshi Ishii at the MIT Media Lab.

2. In formal terms, "skeuomorphism" refers to an element of design or structure that serves little or no purpose in the artifact fashioned from the new material but that was essential to the object made from the original material. One example is the classic Save icon in many computer programs. Typically, this icon shows the form of a physical floppy disk, and while the floppy disk was essential at one time for saving data, it has nothing to do with the wide range of ways for how we store our data nowadays. For an in-depth description of "non-skeuomorphic design," see Gross, Bardzell, and Bardzell (2014).

3. Similar questions have been raised related to understanding something seemingly immaterial through a material lens. Most recently such questions have surfaced in other fields of research where traditional areas such as history, archaeology, and even organizational studies are now increasingly theorized from the viewpoint of materiality (see, for instance, Hedeager 2011; Meskell 2005; Miller 2005; Leonardi, Nardi, and Kallinikos 2012; and Orlikowski 2010). This increasing interest in understanding our world from the viewpoint of materials has further sparked the development of new areas of research that explicitly take materials as a point of departure, and as an object of study—see, for example, the growing interest in material culture studies (e.g., Hicks and Beaudry 2010). Interest in materials and things, and this notion of materiality, now also provide a growing theoretical perspective in a wide range of other fields, stretching from studies in architecture (e.g., Löschke 2016) to the field of political science (e.g., Bennett 2010).

4. Sometimes TUIs (tangible user interfaces) are presented as an alternative to GUIs (graphical user interfaces), in which the distinction between the material and the graphic is central. And, in many cases TUIs are still viewed as a future alternative to contemporary screen-based interaction design (for a discussion on this, see, e.g., Bell and Dourish 2007 and Dourish and Bell 2014).

5. As such, this book is advocacy for an interaction design paradigm that goes well with contemporary ideas of "digital monism," i.e., the idea that our human world is inseparably digital and nondigital, online and offline—or, in obsolete terms, virtual and real. In digital monism, the human world is a digital-centered hybrid environment that tends to form a single (mono-) continuous substance whose name is simply "reality." This is a standpoint that I will return to and elaborate on in chapter 8.

6. The world's first transistor computer was built at the University of Manchester in 1953.

7. Even though computing was a material concern at this point in time it is interesting to notice that when it comes to the paper based punch card it was both a material (and as such an important part of a computing machinery), at the same time as this physical punch card worked as a representation of data (presented in a format that was hard to read for people, but effective that that time for the computer to process).

8. An idea formulated by Alan Turing during the design of the Turing Machine.

9. For an in-depth discussion on the categorical concerns surrounding computer science across its history from the graphical user interface (GUI), via tangible user interfaces (TUIs) to ubiquitous computing, see Robles and Wiberg (2010).

10. As a very rough estimate, I would say that 99.9% of all current user interfaces are graphical and screen-based.

11. This is typical for a range of user interfaces, from video game consoles such as Kinect to pervasive gaming.

12. This phrase was coined by Jakob Nielsen in 1995 in his seminal paper "Ten Usability Heuristics for User Interface Design." It should be noted here that over time the HCI community has not only been interested in user interface usability from a functional perspective, including a focus on cognitive processes, but has also demonstrated a growing interest in user experience design. As a result, we have expanded the scope of usability studies to include a focus on user experiences, user satisfaction, and fun. For further reading about this, see e.g., Wiberg (2003).

13. The "Trash can" symbol in every desktop interface can serve as an illustrative example. And even the "desktop" is a metaphor brought into the design to guide our understanding of the graphical layout of the interface.

14. Again, the "button" is an example of a metaphor used in GUI design.

15. This in direct contrast to "non-skeuomorphic design," which is further discussed in chapter 4.

16. For an in-depth discussion on the importance of developing a vocabulary to talk about computational compositions, see Wiberg and Robles (2010) and Wiberg (2016).

Chapter 2

1. See, for instance, Benyon (2010) and Preece, Sharp, and Rogers (2015) for a thorough description of how metaphor has been used in interaction design and HCI over the last few decades.

2. For some further reading, see Nielsen (1993) and Nielsen (1994).

3. For a detailed description of the notions of "recall" and "recognition" and these two different approaches to design, see Nielsen (1993).

4. *Usability Engineering* by Jakob Nielsen (1993) offers a more in-depth presentation of usability testing.

5. For instance, the usability labs and technologies for usability testing included eye-tracking equipment developed by the company Noldus.

6. For a longer discussion on the notion of learnability in relation to usability, see Nielsen (1996). For a discussion on the spectrum from novice to expert users, see Nielsen (1993).

7. For one such reference, see Löwgren and Stolterman (2004).

8. For a formal definition of the "material turn" in HCI/interaction design, see Robles and Wiberg (2010) and Wiberg and Robles (2010).

9. For a full presentation of the notion of direct manipulation interfaces, see Hutchins, Hollan, and Norman (1985).

10. For a more in-depth presentation of the inFORM shape-shifting display, see Follmer et al. (2013).

11. For more information about the TRANSFORM project, see Ishii et al. (2015).

12. For a further discussion on "radical atoms," see Ishii et al. (2012).

13. Ishii and Ullmer (1997).

14. Oxford Dictionaries, http://www.oxforddictionaries.com/definition/english/manifestation.

15. For a discussion on computing in representational versus presentational form, see Wiberg and Robles (2010).

16. For discussion of design at the intersection between interaction design and architecture, see, for instance, Dalton et al. (2016a); Dalton et al. (2016b); Wiberg (2015a).

17. For further information about this project, see Krassenstein (2015).

18. See, e.g., Vallgårda and Redström (2007).

19. See, e.g., Wiberg (2016) or Vallgårda (2014).

20. See Yao et al. (2015).

21. For some examples of how immaterial materials, including radio, have been brought into composition with computational power, see the work by Sundström et al. (2012).

22. For a discussion on compositional design as an act of bringing elements into compositions and into formal relations, see Robles and Wiberg (2010).

23. For a further discussion on interaction design "beyond the box," see Wiberg (2011b).

Chapter 3

1. On the following pages I will go into some details concerning this question, but for a much longer, and more in-depth discussion on what interaction is and how it can be described and theorized, I recommend the recently published book *The Things That Keep Us Busy: The Elements of Interaction* by Janlert and Stolterman (2017).

2. For one such example of this definition being applied, see Johansson, Skantze, and Gustafson (2014).

3. For an early discussion of this in relation to graphical user interface and more direct forms of interaction, see Sutherland (1963).

4. For a good description of activity theory and how it can be applied in the area of HCI, see Nardi (1995).

5. The term "interactables" was first presented by Mikael Wiberg in a blog post on *ACM Interactions* magazine in 2012 as a one-word notion for interactive objects and objects that might not have embedded computational power but might still be part of a computational process and part of an overall interaction design. This notion has also more recently been further defined and presented in Wiberg (2016) and also elaborated on in Wiberg (2017).

6. See, for instance, Grandjean and Kroemer (1997).

7. This is something that Barad (2007) refers to as "the ontological inseparability of intra-acting agencies."

8. For a full presentation of these ideas, see Barad (2007).

9. See, for instance, Robles and Wiberg (2010), Wiberg (2014), and Wiberg (2015b).

10. Interaction might not be restricted to humans only. For instance, there is interesting ongoing work in the area of HCI / interaction design on animal-computer interaction (see, for instance, Weilenmann and Juhlin 2011).

11. This quote is from Mark Weiser´s classic paper "The Computer for the 21st Century" (Weiser 1991).

12. For an in-depth discussion of "strong concepts" and why they are important for the development of our field, see Höök and Löwgren (2012) and the related paper Stolterman and Wiberg (2010).

13. For an in-depth discussion of how the notion of ubiquitous computing has also served as a vision for development, see the work by Bell and Dourish (2007).

14. For a discussion of the third wave of HCI (and how it differs from first- and second-wave HCI), see Bødker (2006) as well as the follow-up discussion in Bødker (2015).

Chapter 4

1. See, for instance, the following papers published over the last seven years that demonstrate this interest: Robles and Wiberg (2010); Vallgårda and Sokoler (2010a; 2010b); Jung and Stolterman (2011); Gross, Bardzell, and Bardzell (2014); Giaccardi and Karana (2015); and Vallgårda et al. (2016).

2. For a longer discussion on an alternative view of computing, beyond any "metaphorical maneuver" in terms of understanding computing as a material, see Vallgårda and Redström (2007).

3. Some smartphone displays can even distinguish between soft and hard touch and different touch patterns, including distinguishing between different fingerprints.

4. See the work by Donald Norman (1988) for an in-depth presentation of the gulf of execution/gulf of evaluation model.

5. See for instance the design work by Jonathan Ive on non-skeuomorphic interaction design and so-called flat interfaces in the redesign of the interface for iOS running on the iPhone and iPad.

6. See for instance the work by Gross, Bardzell, and Bardzell (2014).

7. In relation to this idea of working with a predefined, well-established, well-known form, we can think about what "form-driven interaction design" could be. For instance, such design could be about *inscribing* interactivity into a familiar form, or as an alternative, we might already know what kind of interaction we want to design, but the design process becomes one of exploring a new and maybe unknown form. In the latter case of expressing interactivity through a new form rather than trying to fit it into an already defined form, I think of this as being an *exscription* of interactivity in material form.

8. A completely different approach to the same use scenario could have been to rely on the sensors in the smartphone and, with that device as a point of departure,

design for similar functionality, but in the form of a visual interface and a smart-phone app.

9. See, for instance, the work by Daniela Rosner on craft-based approaches to inter-action design; for instance, Rosner and Ryokai (2009) and Bardzell, Rosner, and Bardzell (2012).

10. On the importance of understanding materials, and the close relation between how one arrives at this understanding through the hands-on experience of and experimentation with materials during craft sessions, I refer to the work by Donald Schön (1984), which gives a detailed description of how this plays out in practice.

11. For further discussion on digital materials, see the paper "Computational Composites" by Vallgårda and Redström from 2007.

12. The idea that digital materials are increasingly "woven" into the fabrics of our everyday life is quoted from Mark Weiser (1991).

13. For the full article, see Ishii (2008).

14. There is actually a full spectrum of new materials being developed at the moment; for a good overview, see for instance the books on "transmaterials" by Brownell (2006; 2008).

15. For some recent work on shape shifting and shape switching in the area of inter-action design, see for instance the work by Juhlin et al. (2013).

16. For anyone interested in the development of guiding concepts for interaction design and a particular method for doing this, see Stolterman and Wiberg (2010), and for an in-depth discussion on the design of interaction "through a material lens," see Wiberg (2014).

17. Some researchers have even challenged that the user is a he or she, or even a human and has for instance explored not only human-computer interaction, but also animal-computer interaction from one such pespective. As one example see Weilenmann and Juhlin (2011).

18. For some further readings on this topic see: http://www.ign.com/articles /2016/09/07/apple-announces-apple-watch-series-2 and also the discussion here: http://www.theverge.com/2016/9/7/12826994/apple-watch-series-2-hands-on-GPS -waterproof-Nike-ceramic

Chapter 5

1. There are already existing examples of such machine-to-machine interaction between self-driving trucks; a set of self-driving trucks are able to keep the distance between trucks to a minimum to decrease the wind effect and, as a result, save gas.

2. The clarity with which the artifact communicates to its user its changes of state and how it works is a well-researched area in HCI and is typically referred to as "usability studies." For further information, see, e.g., the work by Nielsen (1993; 1994).

3. For a longer discussion on this notion of the disappearing computer, see for instance the book by Norman (1998), *The Invisible Computer*.

4. When I say "user interfaces" here, I refer to the full range of interfaces for human-computer interaction, from traditional graphical user interfaces to interfaces for tangible, embedded, or embodied interaction.

5. For a good, classic discussion on the pros and cons of direct manipulation versus agents, see Shneiderman and Maes (1997).

6. See, for instance, the work by Norman (1988) on the notion of "affordance" and the work by Nielsen (1993) on usability.

7. For an in-depth discussion on agency in the context of human-computer interaction, see Suchman's book *Human-Machine Reconfigurations* (2007).

8. This ranges from simple Excel scripts to IFTTT—"If This Then That"—a new online service that allows users to easily define and configure scripts to automatically control and interact with other digital services and online data (like the stock market or online weather stations) and connect such data with actions (for example, if it rains, send an email or turn on lights), and to configure these scripts to work across different devices, computers, and even embedded systems in smart homes and vehicles.

9. For a longer discussion on how inseparable computing has become from other aspects of our world, see Wiberg (2016).

10. A Uniform Resource Identifier (URI) is a string of characters used to identify a resource, and RDF stands for Resource Description Framework, a family of the World Wide Web Consortium (W3C) specifications.

11. Recently this notion of "screenless interaction" has been referred to as "faceless interaction." For a good introduction to this notion, see Janlert and Stolterman (2015).

12. Compare this with the discussion of an interactive artifact's "dynamic gestalt," which I introduced at the beginning of this chapter.

13. Existing interaction design projects have already explored how wood, and the bending of wood, can be part of a design and be used as a way of expressing the dynamic changes of state. "Planks" (see Vallgårda 2008) is one project that has explored this property of wood in the context of interaction design.

14. It is sometimes claimed that the Sami people have more than 300 words to describe different forms of frozen water, i.e., snow and ice, and studies of the Sami languages in Scandinavia, including studies conducted in Norway, Sweden, and Finland, conclude that the languages have anywhere from 180 snow- and ice-related words and as many as 300 different words for types of snow, tracks in snow, and conditions of the use of snow. For more information about these studies, see Magga (2006) and Jernsletten (1997).

15. In the next chapter I will return to this example and will elaborate on its implications in relation to a material-centered approach to interaction design.

16. For a demonstration of Julius Popp's *bit.fall* water display, see http://vimeo.com /77331036 (December 2015).

17. For further reading on usability, see Nielsen (1993; 1994); for information on contextual inquiry/design, see Holtzblatt et al. (2005); and as a good starting point for reading about the user experience design approach, see Buxton (2007).

18. For a discussion of interaction design as an increasingly relational concern and a matter of crafting meaningful formal relations, see Robles and Wiberg (2010).

19. For a further discussion of this notion, see Löwgren and Stolterman (2004).

Chapter 6

1. See the introduction to this book for my discussion of interaction design as "a relational practice." As a reminder, it is a way of talking about how interaction design is increasingly bringing together different parts into a larger whole that is functional, useful, meaningful, and aesthetically pleasing.

2. For a further theoretical discussion on composition, I recommend the book *Theorizing Composition* edited by Mary Lynch Kennedy (1998).

3. In this chapter I will mostly discuss this particular notion of composition in relation to interaction design. However, for a further discussion of composition and how it relates to design, I strongly recommend Nelson and Stolterman (2012).

4. Jobs is quoted in Isaacson (2011).

5. In this way, the act of composing interaction is quite similar to the design process described by Schön (1984) in that the process is dynamic, dialectic, and a type of "conversation" with the materials at hand.

6. For instance, professional interaction designers often work with code libraries and example codes when programming and developing new interactive artifacts and systems. These code libraries cannot be used in a "cut-and-paste" fashion, so an important task for the interaction designer is to take these "parts," judge how they

might be useful in the composition, and then make small adjustments so that the parts will work in relation to the overall design.

7. This second aspect of compositional interaction design—designing interaction so that it works across the whole composition (beyond only being enabled by it)—is typically referred to as a matter of user interface "consistency" in the HCI literature (see, for instance, Preece, Sharp, and Rogers 2015).

8. See the work by Vallgårda et al. (2016) on material programming.

9. See Robles and Wiberg (2011) for a discussion of the relevance of this particular example for the area of interaction design. And for some further reading about interaction design and architecture, see Wiberg (2010b; 2011a; 2011b; 2012).

10. There is, in fact, a growing interest in the interaction design research community in exploring what we can learn from the field of fashion for the further development of interaction design, and vice versa. For examples of some recent work, see Juhlin et al. (2013).

11. For further reading on this topic, see Baldwin and Clark (2000).

12. This need for new approaches to HCI, including the identified need to integrate theory with HCI design work, has been articulated for more than fifteen years in our field. See, for instance, Sutcliffe (2000).

13. In my view, theory and empirical observations are highly intertwined. Theory should address empirical phenomena, enabling us to see and understand the world around us from a particular viewpoint, but it should also give us a vocabulary to talk more precisely about different objects and aspects of our everyday world. As such, our observations give us the motivation to theorize, and our theories help us to ask further questions about the things we see.

14. This is also a question I have elaborated on further in Wiberg (2010a).

15. For a description and theoretical discussion of how the materiality of interaction emerges and is co-produced during interaction, see Åhman (2016).

16. As such, this double-loop entanglement of the materiality of interaction fits well with other performance-based approaches to interaction design.

17. Siri is the voice-controlled user interface for the iPhone.

18. The development toward "shape-shifting" interfaces is indeed an emergent area of HCI/interaction design. For just one example, see Juhlin et al. (2013).

19. Douglas Engelbart was the first person to talk about text produced in a word processor in terms of materials. He stated that connection on December 9, 1968, in a demonstration video from Augmentation Research Center at Stanford Research Institute in Menlo Park, CA.

Chapter 7

1. As noted in chapter 6, the turn-taking models of interaction were useful for designing command-based user interfaces and agent-based user interfaces (where the user and the computer should "take turns" in the conversation). However, nowadays most interfaces are graphical, and many interaction design models follow more dynamic interaction models that for instance allow for multitasking and easy switches between different tasks, different applications, and different threads of ongoing interaction.

2. See the model of the materiality of interaction as introduced in chapter 6.

3. For a detailed discussion of a concept-driven approach to interaction design, see Stolterman and Wiberg (2010).

4. The notion of "affordance" refers to the perceived functionality of an interactive system. For a more in-depth discussion on the notion of affordance in relation to interaction design, see Norman (1988).

5. See, for instance, Latour (2007).

6. As I discussed in chapter 6, this was a key concern for Steve Jobs. What the design looked like was one concern, but the most important thing was *how* the design works. This concern is a relational concern because it stresses how a certain design should work in relation to someone and in relation to their expectations of how it should work. As such, this relational concern is at the same time a concern for users' experience of a design.

7. See, for instance, Blevis (2007) for an in-depth discussion on sustainability in relation to HCI/interaction design.

Chapter 8

1. "Skeuomorphic design" is any digital design that borrows its properties, look, and feel from similar objects in the physical world.

2. For an in-depth exploration of this idea in relation to both interaction design and our language, which is full of connections to the physical world and how we approach it through a bodily understanding of our everyday experience, see Lund (2003) on the "massification of the intangible."

3. For a further discussion of methodological approaches to reimagining materials in interaction design projects, see Wiberg (2014).

4. For a detailed discussion of this notion of "faceless interaction," see Janlert and Stolterman (2015).

5. Still, it is only from the viewpoint of the user's experience of the digital service that it can be said that cloud-based services are an example of a development that moves away from material concerns. From a design perspective, cloud-based services demand heavy technical infrastructures, high-speed networking, etc., and thus to a great extent are a material concern. For an in-depth discussion, see Dourish (2017).

6. See Simon (1969) for further discussion of how design is about moving from existing situations to future preferable situations.

7. Some people might argue that computing is all about physics. However, as I have argued in this book, the future of interaction design transcends the computer. Beyond the computer, I see interaction design that uses computing as a way of integrating matters across any imaginable substrates—physical or digital, material or immaterial.

Bibliography

Abowd, G., and E. Mynatt. 2000. "Charting Past, Present, and Future Research in Ubiquitous Computing." *ACM Transactions on Computer-Human Interaction* 7(1): 29–58.

Åhman, H. 2016. "Interaction as Existential Practice." Doctoral dissertation, KTH Royal Institute of Technology, Sweden.

Baldwin, C. Y., and K. B. Clark. 2000. *Design Rules: The Power of Modularity*. Cambridge, MA: MIT Press.

Barad, K. 2007. *Meeting the Universe Halfway: Quantum Physics and the Entanglement of Matter and Meaning*. Durham: Duke University Press.

Bardzell, S., D. K. Rosner, and J. Bardzell. 2012. "Crafting Quality in Design: Integrity, Creativity, and Public Sensibility." In *Proceedings of the Designing Interactive Systems Conference (ACM)*, 11–20. New York: ACM Press.

Barnard, P., J. May, D. Duke, and D. Duce. 2000. "Systems, Interactions, and Macrotheory." *ACM Transactions on Computer-Human Interaction* 7(2): 222–262.

Bell, G., and P. Dourish. 2007. "Yesterday's Tomorrows: Notes on Ubiquitous Computing's Dominant Vision." *Personal and Ubiquitous Computing* 11(2): 133–143.

Bellotti, V., M. Back, W. K. Edwards, R. Grinter, A. Henderson, and C. Lopes. 2002. "Making Sense of Sensing Systems: Five Questions for Designers and Researchers." In *Proceedings of the SIGCHI Conference on Human Factors in Computing Systems* (CHI '02), 415–422. New York: ACM Press.

Bennett, J. 2010. *Vibrant Matter: A Political Ecology of Things*. Durham: Duke University Press.

Benyon, D. 2010. 2nd ed. *Designing Interactive Systems*. New York: Addison and Wesley.

Blevis, E. 2007. "Sustainable Interaction Design: Invention and Disposal, Renewal and Reuse." In *Proceedings of the SIGCHI Conference on Human Factors in Computing Systems* (CHI '07), 503–512. New York: ACM Press.

Bødker, S. 2006. "When Second Wave HCI Meets Third Wave Challenges." In *Proceedings of NordiCHI '06—The Nordic Conference on Human-Computer Interaction*, 1–8. New York: ACM Press.

Bødker, S. 2015. "Third-Wave HCI, 10 Years Later: Participation and Sharing." *ACM Interactions* 22(5): 24–31.

Boehner, K., R. DePaula, P. Dourish, and P. Sengers. 2005. "Affect: From Information to Interaction." 4th Decennial Conference on Critical Computing: Between Sense and Sensibility.

Brownell, B. 2006. *Transmaterial: A Catalog of Materials That Redefine Our Physical Environment*. Princeton, NJ: Princeton Architectural Press.

Brownell, B. 2008. *Transmaterial 2: A Catalog of Materials That Redefine Our Physical Environment*. Princeton, NJ: Princeton Architectural Press.

Buchanan, R. 1998. "Branzi's Dilemma: Design in Contemporary Culture." *Design Issues* 14(1): 3–20.

Buxton, B. 2007. *Sketching User Experiences: Getting the Design Right and the Right Design*. San Francisco: Morgan Kaufmann.

Card, S. K., T. P. Moran, and A. Newall. 1983. *The Psychology of Human-Computer Interaction*. Mahwah, NJ: Lawrence Erlbaum Associates.

Carroll, J. M., and W. A. Kellogg. 1989. "Artifact as Theory-nexus: Hermeneutics Meets Theory-based Design." In *Proceedings of the SIGCHI Conference on Human Factors in Computing Systems* (CHI '89), 7–14. New York: ACM Press.

Carroll, J. M., and M. B. Rosson. 1992. "Getting around the Task-Artifact Cycle: How to Make Claims and Design by Scenario." *ACM Transactions on Information Systems* 10(2): 181–212.

Dalsgaard, P., and L. Hansen. 2008. "Performing Perception: Staging Aesthetics of Interaction." *ACM Transactions on Computer-Human Interaction* 15(3): article no. 13.

Dalton, N. S., H. Schnädelbach, T. Varoudis, and M. Wiberg. 2016a. "Architects of Information." *ACM Interactions* 23(4): 62–64.

Dalton, N. S., H. Schnädelbach, and M. Wiberg et al., eds. 2016b. *Architecture and Interaction: Human-Computer Interaction in Time and Place*. Berlin: Springer.

Dam, A. 1997. "Post-WIMP User Interfaces." *Communications of the ACM* 40(2): 63–67.

Davis, M. 2008. "Toto, I've Got a Feeling We're Not in Kansas Anymore." *ACM Interactions* 15(5): 28–34.

Dourish, P. 2015. "Packets, Protocols, and Proximity: The Materialities of Internet Routing." In L. Parks and N. Starosielski, eds., *Signal Traffic: Critical Studies of Media Infrastructures*, 183–204. Champaign: University of Illinois Press.

Dourish, P. 2017. *The Stuff of Bits: An Essay on the Materialities of Information.* Cambridge, MA: MIT Press.

Dourish, G., and P. Bell. 2014. "'Resistance Is Futile': Reading Science Fiction Alongside Ubiquitous Computing." *Personal and Ubiquitous Computing* 18(4): 769–778.

Dourish, P., and V. Bellotti. 1992. "Awareness and Coordination in Shared Workspaces." In *Proceedings of the 1992 ACM Conference on Computer-Supported Cooperative Work*, 107–114. New York: ACM Press.

Dourish, P., and M. Mazmanian. 2013. "Media as Material: Information Representations as Material Foundations for Organizational Practice." In P. R. Carlile, D. Nicolini, A. Langley, and H. Tsoukas, eds., *How Matter Matters: Objects, Artifacts, and Materiality in Organization Studies*, 92–118. Oxford: Oxford University Press.

Dubberly, H., P. Pangaro, and U. Haque. 2009. "What Is Interaction? Are There Different Types?" *ACM Interactions* 16(1): 69–75.

Ellis, C., S. Gibbs, and G. Rein. 1991. "Groupware: Some Issues and Experiences." *Communications of the ACM* 34(1): 39–58.

Fällman, D. 2011. "The New Good: Exploring the Potential of Philosophy of Technology to Contribute to Human-Computer Interaction." In *Proceedings of the SIGCHI Conference on Human Factors in Computing Systems* (CHI '11), 1051–1060. New York: ACM Press.

Follmer, S., D. Leithinger, A. Olwal, A. Hogge, and H. Ishii. 2013. "inFORM: Dynamic Physical Affordances and Constraints through Shape and Object Actuation." In *Proceedings of the 26th Annual ACM Symposium on User Interface Software and Technology* (UIST '13), 417–426. New York: ACM Press.

Giaccardi, E., and E. Karana. 2015. "Foundations of Material Experiences: An Approach to HCI." In *Proceedings of the 33rd Annual ACM Conference on Human Factors in Computing Systems* (CHI '15), 2447–2456. New York: ACM Press.

Goldstein, N. 1989. *Design and Composition.* Englewood Cliffs, NJ: Prentice-Hall.

Grandjean, E., and K. Kroemer. 1997. *Fitting the Task to the Human.* Boca Raton, FL: CRC Press.

Gross, S., S. Bardzell, and J. Bardzell. 2013. "Structures, Forms, and Stuff." Special issue on "Material Interactions." *Personal and Ubiquitous Computing* 18(3).

Gross, S., S. Bardzell, and J. Bardzell. 2014. "Skeu the Evolution: Skeuomorphs, Style, and the Material of Tangible Interactions." In *Proceedings of Tangible and Embodied Interaction (TEI '14)*. New York: ACM Press.

Grudin, J. 1990. "The Computer Reaches Out: The Historical Continuity of Interface Design." In *Proceedings of the ACM Conference on Human Factors in Computing Systems (CHI '90)*, 261–268. New York: ACM Press.

Hedeager, L. 2011. *Iron Age Myth and Materiality: An Archaeology of Scandinavia, AD 400–1000*. London: Routledge.

Heim, S. 2007. *Foundations for Interaction Design*. Reading, MA: Addison-Wesley.

Hicks, D., and M. Beaudry. 2010. *The Oxford Handbook of Material Culture Studies*. Oxford: Oxford University Press.

Hinckley, K., Pierce, J., Sinclair, M., & Horvitz, E. 2000. "Sensing Techniques for Mobile Interaction." In *Proceedings of the 13th Annual ACM Symposium on User Interface Software and Technology* (UIST '00), 91–100. New York: ACM Press.

Hindmarsh, J., M. Fraser, C. Heath, S. Benford, and C. Greenhalgh. 2000. "Object-Focused Interaction in Collaborative Virtual Environments." *ACM Transactions on Computer-Human Interaction* 7(4): 477–509.

Hollan, J., E. Hutchins, and D. Kirsh. 2000. "Distributed Cognition: Toward a New Foundation for Human-Computer Interaction Research." *ACM Transactions on Computer-Human Interaction* 7(2): 174–196.

Holtzblatt, K., J. B. Wendell, and S. Wood. 2005. *Rapid Contextual Design: A How-to Guide to Key Techniques for User-centered Design*. San Francisco: Morgan-Kaufmann.

Höök, K., and J. Löwgren. 2012. "Strong Concepts: Intermediate-level Knowledge in Interaction Design Research." *ACM Transactions on Computer-Human Interaction* 19(3): 23.

Hornecker, E., & Buur, J. 2006. "Getting a Grip on Tangible Interaction: A Framework on Physical Space and Social Interaction." In *Proceedings of the SIGCHI Conference on Human Factors in Computing Systems* (CHI '06), 437–446. New York: ACM Press.

Hutchins, E., J. Hollan, and D. Norman. 1985. "Direct Manipulation Interfaces." *Human-Computer Interaction* 1: 311–338.

Isaacson, W. 2011. *Steve Jobs*. New York: Simon and Schuster.

Ishii, H. 2008. "Tangible Bits: Beyond Pixels." In *Proceedings of the 2nd International Conference on Tangible and Embedded Interaction*, xv–xxv. New York: ACM Press.

Ishii, H., D. Lakatos, L. Bonanni, and J. B. Labrune. 2012. "Radical Atoms: Beyond Tangible Bits, toward Transformable Materials." *ACM Interactions* 19(1): 38–51.

Ishii, H., D. Leithinger, S. Follmer, A. Zoran, P. Schoessler, and J. Counts. 2015. "TRANSFORM: Embodiment of 'Radical Atoms' at Milano Design Week." CHI '15 Extended Abstracts, April 18–23, Seoul, Republic of Korea.

Ishii, H., and B. Ullmer. 1997. "Tangible Bits: Toward Seamless Integration of Interfaces between People, Atoms, and Bits." In *Proceedings of the SIGCHI Conference on Human Factors in Computing Systems* (CHI '97), 234–241. New York: ACM Press.

Jacob, R. 2006. "What Is the Next Generation of Human-Computer Interaction?" In *Proceedings of the Extended Abstracts on Human Factors in Computing* (CHI '06). New York: ACM Press.

Jacob, R., L. Deligiannidis, and S. Morrison. 1999. "A Software Model and Specification Language for Non-WIMP User Interfaces." *ACM Transactions on Computer-Human Interaction* 6(1): 1–46.

Janlert, L.-E., and E. Stolterman. (1997). "The Character of Things." *Design Studies,* 18(3), 297–314.

Janlert, L.-E., and E. Stolterman. 2015. "Faceless Interaction: A Conceptual Examination of the Notion of Interface: Past, Present and Future." *Human-Computer Interaction* 30(6): 507–539.

Janlert, L.-E., and E. Stolterman. 2017. *The Things That Keep Us Busy: The Elements of Interaction.* Cambridge, MA: MIT Press.

Jernsletten, N. 1997. "Sami Traditional Terminology: Professional Terms Concerning Salmon, Reindeer and Snow." In H. Gaski, ed., *Sami Culture in a New Era: The Norwegian Sami Experience.* Karasjok: Davvi Girji.

Johansson, M., Skantze, G., & Gustafson, J. 2014. "Comparison of Human-Human and Human-Robot Turn-Taking Behavior in Multiparty Situated Interaction." In *Proceedings of the Workshop Understanding and Modeling Multiparty, Multimodal Interactions* (UM3I '14), 21–26. New York: ACM Press.

Juhlin, O., Y. Zhang, C. Sundbom, and Y. Fernaeus. 2013. "Fashionable Shape Switching: Explorations in Outfit-centric Design." In *Proceedings of the SIGCHI Conference on Human Factors in Computing Systems* (CHI '13), 1353–1362. New York: ACM Press.

Jung, H., and E. Stolterman. 2011. "Form and Materiality in Interaction Design: A New Approach to HCI." In *Proceedings of the Extended Abstracts on Human Factors in Computing Systems* (CHI'11), 399–408. New York: ACM Press.

Kennedy, M. L., ed. 1998. *Theorizing Composition: A Critical Sourcebook of Theory and Scholarship in Contemporary Composition Studies.* Westport, CT: Greenwood Press.

Klemmer, S., B. Hartmann, and L. Takayama. 2006. "How Bodies Matter: Five Themes for Interaction Design." In *Proceedings of the 6th Conference on Designing Interactive Systems* (DIS '06), 140–149. New York: ACM Press.

Krassenstein, E. 2015. "Amazing Two-String 3D Printed Violin Is Part of Something Special Which Could Revolutionize Music." 3DPrint.com, March 3. http://3dprint.com/48138/3d-printed-2-string-violin/.

Krippendorff, K. 2005. *The Semantic Turn: A New Foundation for Design.* Boca Raton, FL: CRC Press.

Kweon, S., E. Cho, and E. Kim. 2008. "Interactivity Dimension: Media, Contents, and User Perception." In *Proceedings of the 3rd International Conference on Digital Interactive Media in Entertainment and Arts* (DIMEA '08), 265–272. New York: ACM Press.

Latour, B. 2007. *Reassembling the Social.* Oxford: CRC Press, Oxford University Press.

Leonardi, P., B. Nardi, and J. Kallinikos, eds. 2012. *Materiality and Organizing: Social Interaction in a Technological World.* Oxford: Oxford University Press.

Lim, Y.-K., Stolterman, E., Jung, H., & Donaldson, J. 2007. "Interaction Gestalt and the Design of Aesthetic Interactions." In *Proceedings of the 2007 Conference on Designing Pleasurable Products and Interfaces* (DPPI '07), 239–254. New York: ACM Press.

Lim, Y.-K., Stolterman, E., & Tenenberg, J. (2008). "The Anatomy of Prototypes: Prototypes as Filters, Prototypes as Manifestations of Design Ideas." *ACM Transactions on Computer-Human Interaction,* 15(2), 7.

Löschke, S., ed. 2016. *Materiality and Architecture.* London: Routledge.

Löwgren, J., and E. Stolterman. 2004. *Thoughtful Interaction Design.* Cambridge, MA: MIT Press.

Lund, A. 2003. "Massification of the Intangible: An Investigation into Embodied Meaning and Information Visualization." PhD dissertation, Umeå University, Sweden.

Magga, O. H. 2006. "Diversity in Saami Terminology for Reindeer, Snow, and Ice." *International Social Science Journal* 58(187): 25–34.

Marchionini, G., & Sibert, J. (1991). "An Agenda for Human-Computer Interaction: Science and Engineering Serving Human Needs." *SIGCHI Bulletin,* 23(4), 17–32.

Meskell, L. 2005. *Archaeologies of Materiality.* Malden, MA: Blackwell.

Miller, D. 2005. *Materiality (Politics, History, and Culture).* Durham: Duke University Press.

Nardi, B. 1995. *Context and Consciousness: Activity Theory and Human-Computer Interaction.* Cambridge, MA: MIT Press.

Nelson, H., and E. Stolterman. 2012. *The Design Way: Intentional Change in an Unpredictable World.* 2nd ed. Cambridge, MA: MIT Press.

Nielsen, J. 1993. *Usability Engineering.* San Francisco: Morgan Kaufman.

Nielsen, J. 1994. "Heuristic Evaluation." In J. Nielsen and R. L. Mack, eds., *Usability Inspection Methods.* New York: John Wiley and Sons.

Nielsen, J. 1995. "Ten Usability Heuristics for User Interface Design." Nielsen Norman Group 1.1 (1995).

Nielsen, J. (1996). "Usability Metrics: Tracking Interface Improvements." *IEEE Software,* 13(6), 12–13.

Norman, D. 1986. *User Centered System Design: New Perspectives on Human-Computer Interaction.* Boca Raton, FL: CRC Press.

Norman, D. 1988. *The Psychology of Everyday Things.* New York: Basic Books.

Norman, D. 1998. *The Invisible Computer.* Cambridge, MA: MIT Press.

Orlikowski, W. J. 2010. "The Sociomateriality of Organizational Life: Considering Technology in Management Research." *Cambridge Journal of Economics* 34: 125–141.

Pierce, J., and E. Paulos. 2013. "Electric Materialities and Interactive Technology." In *Proceedings of the CHI Conference on Human Factors in Computer Systems* (CHI '13), 119–128. New York: ACM Press.

Pinelle, D., Gutwin, C., & Greenberg, S. (2003). "Task Analysis for Groupware Usability Evaluation: Modeling Shared-Workspace Tasks with the Mechanics of Collaboration." *ACM Transactions on Computer-Human Interaction,* 10(4), 281–311.

Poore, H. R. 1976. *Composition in Art.* New York: Dover.

Preece, J., Sharp, H., & Rogers, Y. (2015). *Interaction Design: Beyond Human-Computer Interaction.* Hoboken, NJ: John Wiley and Sons.

Roberts, I. 2007. *Mastering Composition.* Cincinnati: North Light Books.

Robles, E., and M. Wiberg. 2010. "Texturing the 'Material Turn' in Interaction Design." In *Proceedings of TEI 2011, Fifth International Conference on Tangible, Embedded, and Embodied Interaction,* 137–144. New York: ACM Press.

Robles, E., & Wiberg, M. (2011). "From Materials to Materiality: Thinking of Computing from within an Icehotel." *Interactions (New York, N.Y.),* 18(1), 32–37.

Rosner, D. K., and K. Ryokai. 2009. "Reflections on Craft: Probing the Creative Process of Everyday Knitters." In *Proceedings of Creativity and Cognition '09,* 195–204. New York: ACM Press.

Schön, D. A. 1984. *The Reflective Practitioner: How Professionals Think in Action.* New York: Basic Books.

Shneiderman, B. 1982. "The Future of Interactive Systems and the Emergence of Direct Manipulation." *Behaviour and Information Technology* 1: 237–256.

Shneiderman, B. 1983. "Direct Manipulation: A Step beyond Programming Languages." *IEEE Computer* 16(8): 57–69.

Shneiderman, B., and P. Maes. 1997. "Direct Manipulation vs. Interface Agents." *ACM Interactions* 4(6): 42–61.

Simon, H. A. 1969. *The Sciences of the Artificial.* Cambridge, MA: MIT Press.

Stolterman, E., and M. Wiberg. 2010. "Concept-driven Interaction Design Research." *Human-Computer Interaction* 25(2): 95–118.

Stolterman, E., & Wiberg, M. 2015. "Modeling the Flows of Interactivity." Workshop paper, CHI '15, Seoul, Korea.

Suchman, L. 1987. *Plans and Situated Actions: The Problem of Human-Machine Communication.* New York: Cambridge University Press.

Suchman, L. 1995. "Making Work Visible." *Communications of the ACM* 38(9): 56–64.

Suchman, L. 2007. *Human-Machine Reconfigurations: Plans and Situated Actions,* 2nd ed. Cambridge: Cambridge University Press.

Sundström, P., et al. 2012. "Immaterial Materials: Designing with Radio." In *Proceedings of TEI '12: The Sixth International Conference on Tangible, Embedded and Embodied Interaction,* 205–212. New York: ACM Press.

Sutcliffe, A. (2000). "On the Effective Use and Reuse of HCI Knowledge." *ACM Transactions on Computer-Human Interaction,* 7(2), 197221.

Sutherland, I. 1963. "Sketchpad: A Man-Machine Graphical Communication System." In *Proceedings of AFIPS '63 Spring Joint Computer Conference,* 329–346. New York: ACM Press.

Toney, A., B. Mulley, B. Thomas, and W. Piekarski. 2003. "Social Weight: Designing to Minimize the Social Consequences Arising from Technology Use by the Mobile Professional." *Personal and Ubiquitous Computing* 7(5): 309–320.

Turk, V. 2014. "This Gadget Turns Any Object into Electronic Music." *Motherboard,* February 19. https://motherboard.vice.com/en_us/article/this-gadget-turns-any-object-into-electronic-music.

Vallgårda, A. 2008. "PLANKS: A Computational Composite." In *Proceedings of the 5th Nordic Conference on Human-Computer Interaction* (NordiCHI '08), 569–574. New York: ACM Press.

Vallgårda, A. 2014. "Giving Form to Computational Things: Developing a Practice of Interaction Design." *Personal and Ubiquitous Computing* 18(3): 577–592.

Vallgårda, A., L. Boer, V. Tsaknaki, and D. Svanaes. 2016. "Material Programming: A New Interaction Design Practice." In *Proceedings of the 2016 ACM Conference on Designing Interactive Systems* (DIS '16), 149–152. New York: ACM Press.

Vallgårda, A., and J. Redström. 2007. "Computational Composites." In *Proceedings of the SIGCHI Conference on Human Factors in Computing Systems* (CHI '07), 513–522. New York: ACM Press.

Vallgårda, A., and T. Sokoler. 2010a. "Material Computing: Computing Materials." In UbiComp '10 Adjunct: Proceedings of the 12th ACM International Conference Adjunct Papers on Ubiquitous Computing, Copenhagen, Denmark.

Vallgårda, A., and T. Sokoler. 2010b. "A Material Strategy: Exploring Material Properties of Computers." *International Journal of Design*, 4(3): 1–14.

Vallgårda, A., M. Winther, N. Mørch, and E. Vizer. 2015. "Temporal Form in Interaction Design." *International Journal of Design* 9(3): 1–15.

Waibel, A., & Stiefelhagen, R. (Eds.). (2009). *Computers in the Human Interaction Loop. Human-Computer Interaction Series.* Berlin: Springer.

Weilenmann, A., and O. Juhlin. 2011. "Understanding People and Animals: The Use of a Positioning System in Ordinary Human-Canine Interaction." In *Proceedings of the SIGCHI Conference on Human Factors in Computing Systems* (CHI '11), 2631–2640. New York: ACM Press.

Weiser, M. 1991. "The Computer for the 21st Century." *Scientific American* 26(3): 94–101.

Whittaker, S. 1996. "Talking to Strangers: An Evaluation of the Factors Affecting Electronic Collaboration." In *Proceedings of the 1996 ACM Conference on Computer-Supported Cooperative Work* (CSCW '96), 409–418, New York: ACM Press.

Whittaker, S., J. Swanson, J. Kucan, and C. Sidner. 1997. "TeleNotes: Managing Lightweight Interactions in the Desktop." *ACM Transactions on Computer-Human Interaction* 4(2):137–168.

Whittaker, S., L. Terveen, and B. Nardi. 2000. "Let's Stop Pushing the Envelope and Start Addressing It: A Reference Task Agenda for HCI." *Human-Computer Interaction* 15: 75–106.

Wiberg, M. 2001. "RoamWare: An Integrated Architecture for Seamless Interaction in between Mobile Meetings." In *Proceedings of the 2001 International ACM SIGGROUP Conference on Supporting Group Work*, 288–297. New York: ACM Press.

Wiberg, C. 2003. "A Measure of Fun: Extending the Scope of Web Usability." PhD thesis, Department of Informatics, Umeå University, Sweden.

Wiberg, M. 2010a. "Interaction per se: Understanding 'the Ambience of Interaction' as Manifested and Situated in Everyday and Ubiquitous IT-use." *International Journal of Ambient Computing and Intelligence* 2(2): 1–26.

Wiberg, M. 2010b. "Interactive Architecture as Digital Texturation: Transformed Public Spaces and New Material Integration." In J. Holmström, M. Wiberg, and A. Lund, eds., *Industrial Informatics: Design, Use and Innovation*, 44–57. Hershey, PA: IGI Global.

Wiberg, M. 2011a. *Interactive Textures for Architecture and Landscaping—Digital Elements and Technologies, Information Science Reference*. Hershey, PA: IGI Global.

Wiberg, M. 2011b. "Making the Case for 'Architectural Informatics': A New Research Horizon for Ambient Computing?" *International Journal of Ambient Computing and Intelligence* 3(3): 1–7.

Wiberg, M. 2012. "Landscapes, Long Tails and Digital Materialities: Implications for Mobile HCI Research." *International Journal of Mobile Human Computer Interaction* 4(1): 45–62.

Wiberg, M. 2014. "Methodology for Materiality." *Interaction Design Research through a Material Lens: Personal and Ubiquitous Computing* 18(3): 625–636.

Wiberg, M. (2015a). "Interaction Design Meets Architectural Thinking." *Interactions (New York, N.Y.)*, 22(2), 60–63.

Wiberg, M. 2015b. "Interactivity and Material Experiences: A Literature Study on Materiality." *Making and Thinking* 1(4).

Wiberg, M. (2016). "Interaction, New Materials and Computing: Beyond the Disappearing Computer, towards Material Interactions." *Materials & Design*, 90(15), 1200–1206.

Wiberg, M. (2017). "From Interactables to Architectonic Interaction." *Interactions (New York, N.Y.)*, 24(2), 62–65.

Wiberg, M., and E. Robles. 2010. "Computational Compositions: Aesthetics, Materials, and Interaction Design." *International Journal of Design* 4(2): 65–76.

Yao, L., J. Ou, C.-Y. Cheng, H. Steiner, W. Wang, G. Wang, and H. Ishii. 2015. "bioLogic: Natto Cells as Nanoactuators for Shape Changing Interfaces." In *Proceedings of the 33rd Annual ACM Conference on Human Factors in Computing Systems* (CHI '15), 1–10. New York: ACM Press.

Index